Happy Plant

Happy Plant

A Beginner's Guide to Cultivating Healthy Plant Care Habits

PUNEET SABHARWAL

RESEARCH BY **Christopher Satch**

ILLUSTRATIONS BY **Travis DeMello**   PHOTOGRAPHS BY **Cayla Zahoran**

Princeton Architectural Press · New York

To my mom, for forcing me to water her plants.
To the Brooklyn Beloveds, for believing in Horti.
To all the plants I have unintentionally killed—
this book wouldn't exist without you.

CONTENTS

FOREWORD

There are plenty of books out there for houseplant enthusiasts to dig into (we know because we wrote one!), but we're glad you've chosen *Happy Plant*. This is a book that will share the fascinating facts of plant evolution as well as essential care advice, all told in in Puneet Sabharwal's signature encouraging voice. We met Puneet through Instagram and quickly learned that he is a wise and giving plant person who shares our love for all things green and hopes to share this love with as many people as possible.

Reading this book is like stepping into Puneet and Bryana Sortino's immersive studio called Horti Play in Greenpoint, Brooklyn, where plants are treated with respect and admiration. In these pages, you will learn more about why plants deserve such high praise as well as the basic care tips that will keep them thriving in your home.

Whether you've picked up this book in order to care for your first plant or your hundredth, Puneet will take you step-by-step through how to manage that special relationship in a straightforward and nonjudgmental way. Once you begin bringing plants into your home, it can be hard to stop. However, taking care of plants becomes monumentally easier with a commonsense guide like this one. It's a thrill to have a houseplant thrive under your care, and so we love the honest approach behind *Happy Plant*.

As your knowledge grows, so will your plants. We hope you take your time reading these pages, and study your own green beauties based on what you learn here. For plant rookies, pick out a plant from Puneet's list of best first plants (see page 67; our favorite is the snake plant), or if you're looking for a new plant pal to bring home, check out the guide to common houseplants (we like *Pilea peperomioides* on page 141). Over time, your plant IQ will increase as you interact with more and more foliage friends. We hope that each and every one is a Happy Plant!

Morgan Doane & Erin Harding of
@houseplantclub

INTRODUCTION

During the 1980s, I grew up in a commune in India, where my mom cared for an abundance of plants in our sun-drenched courtyard. Given that temperatures in Delhi can reach 115°F (47°C), the plants had to be watered every day. Watering the plants was a chore, a duty assigned to me that disrupted my day. In the evenings, after I had finished my homework, I had to water all of the plants before I was free. This was my first relationship with houseplants, and I hated them. Plants were killjoys that ate up thirty minutes of playtime. Fast-forward two decades. I immigrated to the United States, and what was the first item I bought for my apartment? A houseplant.

In the ensuing years, I acquired more than seventy plants and designed every aspect of my apartment around these living creatures. Not only that, I launched a company, Horti, that teaches others how to build meaningful, lasting relationships with houseplants. Plants have influenced my aesthetic and design taste, my work, my home, my inner well-being, even my dating life—which eventually led me to my business partner Bryana Sortino. For me, plants provided a sense of rootedness when I left my home country and moved to the other side of the world. The little garden I hated when I was a kid in India became my safe haven in the US, a connection that had been ingrained in me by my mother. I do not have formal training in horticulture, but I do share an emotional connection with plants. I may not remember all the complex Latin names of many plants, but I am learning to pay attention to their songs.[1]

If you are totally new to plant care, I am here to let you know that it does not matter if you can pronounce a plant's Latin name or if you can spot overwatered soil. The reality is that most of us have killed more houseplants than we care to admit, and there is no such thing as a green thumb. Just like love, plant care is not a magic potion, it is a dialogue.

When you bring home a new houseplant and give it a name, it is a little bit like bringing home a newborn, except, too often, the only instructions for your plant's survival are a few words printed on the identification card tucked into its soft soil. The initial joy of acquiring a plant can quickly turn into panic, especially if it is your first houseplant. Unlike babies, houseplants do not offer cues to parenting but only sit silently, watching, feeling, and slowly responding to the effects of human trial-and-error care.

This is my effort to highlight all that I have learned on my own plant-parenting journey. I hope this book can serve as a companion or coparent to help you succeed in your new phase of life with plants. Outlining the different stages of your plant's development, I will explain how to learn the language of your plant and nourish its growth by cultivating healthy plant-parenting habits. And, as equally important, this book offers insight into what makes a plant a houseplant and how exactly they find their way to you.

Your relationship with plants is personal, and the real secret to building a sustainable connection with houseplants is to recognize the potential for mutual dependence between humans and nature instead of assuming human dominance over nature. My own relationship with plants—and with caring for plants—changed when I realized that, in reality, they were taking care of me.

—Puneet Sabharwal

Chapter One: Plants

WHERE DO PLANTS COME FROM?

I might have been six or seven years old when I cut into the stem of a rubber plant (*Ficus elastica*, 'Ruby') in the verandah of my childhood home and watched, strangely mesmerized, as the white fluid bled out of the plant. It was common to see my mom clip a leaf or two of tulsi, or holy basil, and boil the leaves in her chai for flavor. The idea that plants are alive never occurred to me until one day, soon after creating a small stab in my favorite victim, I went inside and saw an ad playing on TV about deforestation. In it, an animated plant talks about feeling pain, encouraging people to stop cutting down trees and to plant more seeds for the future. I went outside with tears dripping down my face and apologized many times to the plant before placing a Band-Aid over its wounds. It would take me many years to fully understand how humans develop a relationship of mutual care with nature and that plants are intelligent creatures. The one positive from that experience was that I did learn that they are truly living.

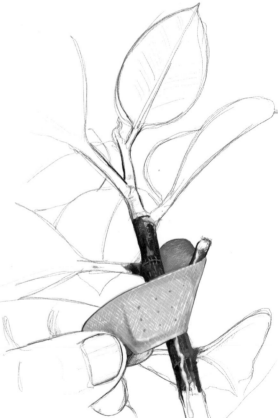

Six-year-old me making a small cut into my mother's rubber plant to watch the white fluid bleed out of its stem before applying a Band-Aid to the wound.

Have you ever thought about how the language we use to talk about plants parallels how we discuss our own human anatomy and bodily functions? Just look at how the veins of leaves resemble the veins in your own hands. Every one of the 350,000 known plant species experiences the same biological processes as organisms, microbes, and animals like us. Plants represent a wide range of colors, shapes, and sizes and serve as sources of inspiration for human creativity and innovation. Their cell walls provide structure to our material lives in the form of plant fibers spun into clothes; wood used to build structures and burn for energy; and paper on which to compose music, make art, and print books like the one in your hand.

While we still often categorize plants as decorative objects in our homes, the emerging field of plant neurobiology is helping us understand how perceptive plants are to their surroundings. Plant neurobiology is founded on the belief that plants are fundamentally intelligent life-forms that can communicate with each other, navigate their environments, and solve problems.[1] Take single-celled, or unicellular, organisms like algae, for example, which exhibit responses to stimuli and make choices to escape harm or pursue pleasure. If they can possess complex behaviors, then plants, which are composed of billions of such organisms, should, at the very least, have the same potential.

Outside of the realm of conventional science, there are many Indigenous traditions that consider plants as perceptive beings. The Kutia Kondh people of the state of Odisha in India perceive all plants, from canopies of trees to plains of prairie grass, to be social creatures with their own respective communities that communicate and interact with one another.[2]

Each plant possesses individual characteristics and behaviors that collectively compose green landscapes and underwater forests. Plants are the foundational food producers of Earth's ecosystems, providing shelter, clean water, and oxygen for neighboring organisms. Notice the way that air courses through your nose when you take a breath, filling your lungs and belly, and gently exiting your mouth. Like you, plants respire, and they thrive on sunlight and water. There are multitudes of genes, hormones, and sexualities that are contained within all plants. They reproduce and reinvent themselves—adapting in uncertainty, protecting themselves in danger, sending help to sickly or injured members of their communities, and flourishing in healthy environments.

1.1
A Short History of Plant Evolution

The world of plants is fascinating, but the information can be a bit murky and is too often explained in clunky, technical terms. When I first began my plant subscription company, Horti, I found it difficult to access practical information about the origins of plants and their inner workings. While writing this guide, I consulted with plant scientist and researcher Christopher Satch to help break down plant history and biology, and to shed light on how houseplants became an important part of our lives.

More than half a billion years ago, the first plants came to land from the sea: algae, in the form of floppy kelp, would wash ashore and tolerate desiccation just long enough for the tides to come in and resuscitate them with life-giving water. But they did not do this alone. The story of how plants evolved to survive on land would not have been possible without help from fungi and bacteria.

There are countless fungi and bacteria, with some bacteria, such as cyanobacteria, specializing in harvesting nutrients by forming a film on rocks near water. The film slowly degrades the rock, allowing the bacteria to gather

Early Earth, around five million years ago, is thought to have looked much like modern-day Iceland—covered in mosses and other small primitive plants.

minerals for use in photosynthesis to feed themselves. (Photosynthesis is a process that cyanobacteria as well as plants experience, albeit in different forms.) Also known as blue-green algae, cyanobacteria are ancient organisms that are the first known to conduct photosynthesis. As a result of scarce minerals to feed upon, or of drying out, bacteria die and are then fed upon by fungi, creating a more powerful acid to dissolve the rock, with live bacteria supplying the sugars in exchange for some of the nutrients. The biofilm kept these kelp-like creatures, or plant ancestors, wet enough to survive until the tide came back in, and eventually, the plantcestor replaced the photosynthetic bacteria by doing the work of providing fungus with sugar—a primitive sugar daddy, if you will.

The interdependent and mutually beneficial relationship between ancient plants and fungi became successful

PLANTS

Lichens are a symbiotic association between a fungus and algae, cyanobacteria, or both. The algae photosynthesize to provide sugars and the fungus provides moisture and minerals.

enough to intertwine their destinies for the rest of time. Early Earth, once rocky before life appeared, became covered with mosses, liverworts, and other primitive plants that had no vascular systems. While land was covered in moss, these early plants were unable to grow past a certain height, and, for a time, no vegetation existed on Earth that was more than two feet tall. Lichens, ancient composite organisms, are an interesting evolutionary relic from an earlier era that persist to this day, unchanged.

Over time, plants evolved vascular systems, which allowed them to grow taller and live farther from the sea. By this time, over 420 million years ago, enough genera-tions of moss had lived and died to add organic matter made up of carbon compounds to the earth, creating the first soil. Young Earth closely resembled Iceland today, with vast mounds of moss, rocks, and lichens. It was in this damp, rainy environment that protoplants could live inland.

Humans and a majority of plants have similar vascular systems—both are made up of tubes that transport metabo-lites, water, and waste. In humans, our circulatory system is our vascular system, and our veins and arteries do the work of transporting fluids to and from the heart. In plants, these systems are called the xylem and the phloem. The xylem brings nutrients and water from the soil up to the leaves, and the phloem brings the sugars down to feed the roots, which cannot make their own food. Fungi form associations with the roots to extend the surface area to maximize their intake of minerals and nutrients.

All nonvascular plants have the same basic anatomy: water is pulled up from the roots by the drying power of the sun and released through the leaves. While nearly all plants take in water through their roots, some nonvascular plants have special structures that allow them to absorb water through their leaves; these include mosses and

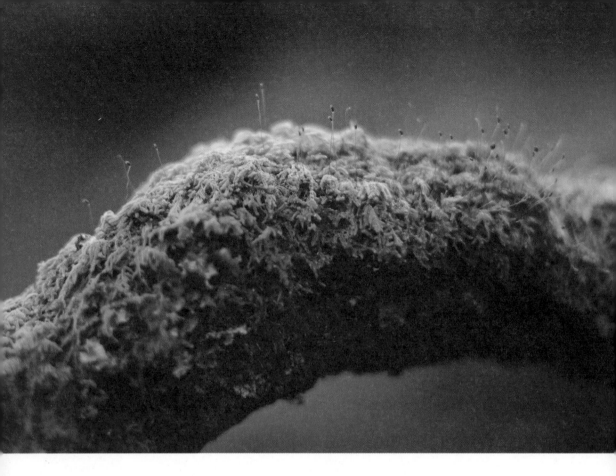

Mosses are generally the first visible organisms to colonize a new location, assuming there is enough water to sustain life. As mosses grow upward, the lower, older parts die, decompose, and become humus, the organic component of soil.

certain epiphytes (plants that grow on the shallow surface of other plants, like orchids) and bromeliads (tropical plants) with short stems and spiny leaves. But even among these epiphytic plants, the majority of their water intake is still consumed through the roots.

Plants need soil to soak up water. Long ago, organic matter from generations of decomposed mosses helped soil retain water after a rainstorm, which allowed the primitive roots of plants to evolve to absorb the water. Over time, plants began to grow taller and compete with each other for sunlight, the source of all energy on the planet. Once plant life was established and thriving on Earth, animal life followed.

Fig. 126. A, reconstruction of *Lepidodendron elevatum* Sternb. B, reconstruction of a species of *Sigillaria*. (After Hirmer, 1927.)

An artist's rendition of what lycopods may have looked like, based upon fossil evidence. Before the Permian-Triassic extinction, life on Earth looked radically different from how it appears today.

Staying alive on Earth has never been easy, no matter what species you are or what millennia you find yourself in. Various extinctions have eliminated certain plant species from the face of Earth forever. Take the Permian-Triassic extinction, Earth's most severe extinction, colloquially known as the Great Dying, during which more than 80 percent of all genera, or all groups of species, became extinct. During this period, plant life was severely reduced by toxic volcanic eruptions, acid rain, abrupt desertification, and ice ages. When Earth stabilized, new plant microenvironments developed greater complexities and created niche roles that emerging plants evolved to fill. There was an evolutionary explosion of diversity, and what we now think of as modern plants emerged to replace the lycopods and tree-ferns that once dominated forest landscapes before the Permian-Triassic extinction, 250 million years ago.

Gymnosperms, such as cone-producing conifer trees, took over as the dominant plant life in Earth's generally dry climate where they acclimated to both hot and cold temperatures. As Earth became more stable, dinosaurs began to roam the land, and eventually, flowering plants bloomed—that is right, dinosaurs existed on Earth before flowers did! In fact, flowering plants are the youngest in plant evolution, showing up around 140 million years ago—they are quite literally late bloomers. Flowers changed how plants shuffled their genes, which led to another burst of biodiversity, including Earth's first fruits, which only grow from flowering plants and trees.

It may come as a surprise that the majority of houseplants are actually flowering plants too. Just because you have not seen your houseplants bloom does not mean that they are not capable of doing so. Some just have not met the conditions necessary for them to unfurl their blooms. Flowers are the reproductive parts of the plants, and when

Plants have evolved many
mechanisms in order to
reproduce. Flowers are
evolution's latest creation
for plant reproduction.

flowers fall off of a plant, it does not mean that the plant is dying but rather that it is finished feeling sexy. For some plants, it is rare for them to ever be "in the mood." *Pilea*, *Monstera*, and *Dracaena* will flower at some point in their lives but only if the conditions are right.

The popular *Monstera deliciosa*, for example, needs to be fully mature before it flowers and grows fruit, which requires partial sun exposure to trigger the blooms. We often think of houseplants as only being leafy and green, but with consistent tender care, they just might surprise you with flowers one day.

Now that we can grasp where plants are coming from—as soft protokelp that washed ashore hundreds of millions of years ago to evolve into the lush flowering jungles and towering forests that keep our atmosphere intact today—we can better understand what it is that plants need to thrive and how they benefit our lives. We can begin to reimagine our shared resilience and mutual dependence. After all, life on Earth is only possible because of plants.

1.2
Plant Quirks and Understanding Them

To really know a plant, we need to know the role they play in larger ecosystems and the mechanisms that determine their growth patterns and life cycles. The anatomy of a plant is made up of five main organs: roots, stems, leaves, buds, and reproductive structures.

Occasionally, one or more of the plant's organs may be modified by evolution to serve another purpose. For example, in cacti, the leaves are modified to become spines and the stem becomes a giant photosynthetic storage unit to hold water. In calathea plants and irises (as well as many other plants), the stems grow horizontally and are hidden partially or entirely underground. These underground stems are called rhizomes and allow the plant to spread

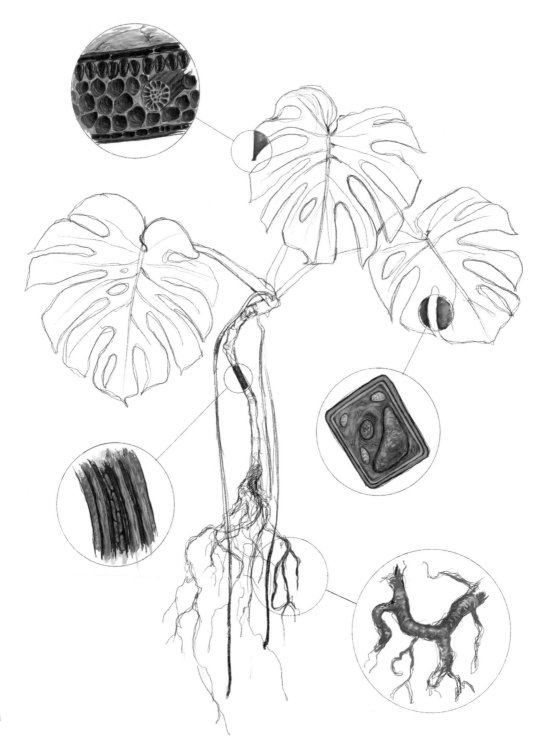

horizontally rather than vertically. If a plant organ appears to be missing, it has likely been altered into something else or has been reduced due to lack of use.

All plants draw water from their roots, which flows through the plants and then evaporates in a process referred to as transpiration. Water is released through the plant's pores called stomata. It is through the upward force of transpiration that minerals (dissolved in water) are pulled directly from the ground through the xylem tubes of the roots providing nutrition to the growing plant. In exchange, the sap produced by photosynthesis in the leaves and in other green parts of the plant flows downward into the neighboring phloem tubes. This is how the root cells are fed, as they cannot make their own food underground. Together, the xylem and phloem make up the circulatory system of the plant.

Have you ever wondered why plants droop or how they manage to hold themselves up without any additional support? The answer is water pressure. Plant cells are like big water balloons. When a plant cell is full of water, it is plump and turgid, forming a solid but pliable block. Each plant cell has a large storage component called a vacuole, which stores and pressurizes the water. However, like humans and animals, plants will become dehydrated if they lose more water than they ingest.

Water deficiency occurs when there is not enough groundwater to replace the water lost during transpiration. To remedy this, water is pulled from the vacuole, where water is stored, and the cells deflate, causing the plant to droop. Oftentimes, you may even see yellow leaves accompanying dehydration or other maladies. Contrary to popular belief, yellow leaves are not just a symptom of overwatering but also a general sign that the plant is under stress (see page 127). If your plant cannot support itself and

Plants absorb nearly all of their water through their roots. Many herbaceous plants hold themselves up with water pressure within the plant cells. Lack of water, and therefore deflation of the cells, causes the plant to droop.

Calatheas are called prayer plants because of the daily upright, or prostrate, positioning of their leaves, which open and close in accordance with their circadian clock.

its new shoots with adequate sunlight, there are likely issues underneath the soil.

Much like how ancient plants were dependent on bacteria and fungi in order to evolve, plants continue to form relationships with other species around them in order to meet their needs. *Hoya* plants and orchids produce a sugary sap from extrafloral nectaries in order to attract ants, who will defend the plants from herbivorous insects. Some plants possess surprising quirks, with morning glory vines and some types of orchid opening and closing their flowers in accordance with circadian cycles, a natural response to light and dark over a twenty-four-hour period. *Calathea* and *Maranta* prayer plants lift and lower their leaves in response to circadian cycles as well.

Many cultivars of houseplants mutate and create genetically different variations of the original plant, which are called sports. While it is unclear what triggers the random formation of sports, each plant propagation is a roll of the genetic dice, and sometimes you hit a lucky lotto number and get variegation, a random coloration of the leaves that is generated from the organelles of plant cells called plastids. True variegation comes from mutated plastids, or little cellular organelles, that create distinct unrepeatable "fingerprint" patterns, as seen in the *Philodendron* Pink Princess and *Philodendron* Snow Queen hibiscus. Variegation can be easily confused with coloration, which comes from the instructions of the nuclear genes of the plant and results in predictable, uniform patterns. The genera of *Calathea* and *Maranta* prayer plants are perfect examples of coloration.

Familiarizing yourself with plant anatomy, colors, and patterns is an important part of understanding your plant and seeing all plants with fresh eyes. In the next section, I dig into the origins of houseplants, which date back thousands of years to when humans first started putting plants into pots and bringing them indoors.

1.3
What Makes a Plant a Houseplant?

For millions of years, plants have grown outdoors, evolving, adapting, and reproducing wherever they could survive. When human civilizations emerged onto the scene approximately ten thousand to twelve thousand years ago, they began to harness the power of plants in a way no other animals had before—through cultivation.[3] No longer did humans need to solely forage and hunt for food, they could grow their own food in one place.[4]

While there was limited communication across geographic divides, and botanical innovations rose and fell with many civilizations, we know that the Sumerians and ancient Egyptians began growing trees and other plants in large vessels around four thousand years ago, while in China, potted ornamental plants had already been grown in courtyards for centuries.[5] It was not until the dawn of the

Keeping ornamental plants indoors has only become possible with recent understanding of plant biology, and is due, as well, to the adaptability of various tropical plants to indoor conditions.

eighteenth century that potted plants finally made their debut indoors in European societies.[6]

In 1652, Sir Hugh Platt, an English horticultural authority, published *The Garden of Eden*, in which he theorized that it was possible to grow plants indoors.[7] Previous attempts had failed, largely because people did not fully understand the quantity of light plants need in order to live, nor did they realize the extent of the differences in the quality of outdoor and indoor light. Platt suggested that plants needed to be placed in a sunny window to survive and that direct sunlight was instrumental to the survival of plants indoors. While glass offered light, the windows of the time did not have modern insulation, creating drafty

A few plants waiting to be watered in the sink. When watering, use warm water and let the soil become completely saturated. Allow the plant to drip dry in the sink before returning the plant to its ideal indoor location.

homes. With that in mind (as well as limited space and even more limited access to luxury materials such as glass and metal), people constructed large, heated glass and metal or stone structures, which were predecessors to the modern greenhouse, rather than focus on growing plants inside of their homes. As a result, houseplants belonged to the elites, who could afford to erect conservatories for plants, which also doubled as symbols of their wealth.

As Western societies encroached on the rest of the world through travel, colonization, and violent conquest, any so-called exotic plants that carried economic or symbolic value were abducted and brought back to be propagated in the collections of the ruling class. Possessing tropical plants and succeeding in keeping them alive throughout the harsh European winter became a status symbol. Centralized heating systems became common in the nineteenth century, which made houses hospitable enough year-round for indoor plants to survive. During the Victorian era, the invention of the Wardian case, an early version of the glass terrarium, made it easier to keep ferns and moisture-sensitive tropical plants safe from the polluted and dry air surrounding them. These smaller glass structures were more affordable than greenhouses, and middle-class people quickly "caught" fern-mania and orchidelirium. By the 1850s, the houseplant industry was well established, and Kew Gardens was the center of the horticultural universe.[8]

The popularity of houseplants continued to fluctuate and nearly collapsed during both world wars, but economic upswings have brought houseplants back into the mainstream. Today, growing indoor plants is accessible to nearly everyone and can be as easy as propagating a plant cutting and waiting for it to grow roots.

In reality, an indoor plant is any plant that can adapt to indoor conditions or any plant whose native environment can be recreated indoors. What makes an ideal home for plants? The most important factor is light. No matter how bright your home may look to you, it is actually darker than your eyes can register. That is because the human eye adjusts to different levels of brightness in order to maintain clarity (see page 88).

Most of the houseplants being sold on the market are tropical understory plants that grow underneath the shaded canopy of trees and therefore do not require as much direct sunlight (think of aroids like the peace lily). Folks who are lucky enough to have sun-drenched south-facing windows—or, if south of the equator, north-facing windows—can provide an indoor climate where light-loving tropical plants and even succulents and cacti can thrive. Succulents and cacti are perfect for windows that get so much direct sun, the glass itself heats up from a steady stream of the sun's rays.

While you cannot bring each and every plant from the outside in, you can focus on the countless houseplants that *do* thrive inside with the right amount of sunlight, water, stable temperatures, and tender loving care.

1.4
Plant Nurseries and Distribution

The first time that I visited a nursery, I was in awe of the lush, happy plants covering every square inch of space available—this was my green heaven. By that time in my life, I had already been spending a considerable amount of time (and money) on my plant collection, but entering into this space strengthened my love for plants and motivated me to pursue a career in plant care. My business partner, Bryana, and I started looking forward to our nursery day when we would refresh our inventory of plants for our company, Horti. Nursery trips became daylong adventures as we wandered through each greenhouse, touching and feeling every species of plant that we came across.

Over the years, we have learned how much attentive care it takes from a network of growers to raise seedlings into plants that find their way to you. The business of growing plants requires patience, even with fast-growing technology, innovations in lighting, the mitigation of diseases, and cultivation through tissue cultures. It is unsurprising, then, that the plant business attracts some of the calmest and most patient people I have ever met. To better understand the roles of local nurseries and plant retailers in bringing houseplants to you, I enlisted the expertise of Jeff Keil, whose family has been growing and selling plants since the 1940s, and Al Newsom, who works

PLANTS

at Ball Horticulture, a global company that has been around for more than a hundred years.

The origin of plant nurseries in the United States can be traced back to 1737, when Robert Prince opened the Linnaean Botanic Garden and Nurseries in Flushing, New York.[9] During the 1720s, Prince wrote letters to sea captains, asking for their help in bringing new plants from abroad. It was not until more than a hundred years later that the field of horticulture was legitimized, when President Lincoln signed a law establishing the United States Department of Agriculture, or as he once called it, "the People's Department."[10]

Plants in a greenhouse are arranged according to their light needs. The plants you buy in stores are often supercharged and extra leafy from all the extra light they receive in the greenhouse.

Meticulous record keeping and rotation of stock is all a part of a day's work for Jorge M. Sisalima at Fantastic Gardens in Long Island, New York.

The twentieth century brought with it a growing interest in ornamental horticulture through gardening clubs. Unlike the formal horticulture societies that preceded them, gardening clubs were open to everyone and provided a relaxed social setting for people to share their interests in growing plants. Plant care was easier than ever with the invention of pesticides, and the invention of double polyethylene greenhouse covers in the 1970s increased the heat retention of greenhouses while lowering their costs, which made year-round plant growing more accessible, especially to people looking to turn their green thumbs into a business.

Today's plant nurseries vary—some sell directly to consumers and others serve as a hub, holding large stocks of plants that have been raised by growers across the country. Nurseries that do the work of growing their own plants typically sell them at wholesale to retailers like Horti. You may think that plant production starts with seeds, but most houseplants are grown through asexual propagation techniques, such as cuttings, grafting, and tissue culture.

For hundreds of years, growers have used cuttings and grafting to mass-produce plants, but there is a limit to what you can grow with a single parent plant. Tissue culture is the most advanced technique and is currently the fastest method, which allows growers to clone plants in a fraction of the time (compared to the years needed to generate mature plants) by removing a small sample of a tissue from a healthy plant and tricking it into multiplying itself, creating an exponential number of new plants.

Growers are like most plant enthusiasts—they love to find eccentricities in plants, like white leaves on a *Monstera* or pink leaves in a batch of *Philodendron* plants. Uncommon markings and coloring in plants are almost impossible to cultivate through seeds. Tissue culture allows growers to breed for unique coloring and variegation, and breeding new varieties of plants leads to new plant names.

The taxonomy of plants is currently made up of standard plants as found in nature and patented plants that have been crossbred or developed in labs by licensed growers. Many people only recognize plants by their common names, which is perfectly fine, but learning the scientific (or Latin) names of plants makes it easier to be specific about an exact type of plant and to communicate about it universally. For example, if you are looking for a money tree plant, you may be surprised to find there are a handful of different plants with the same name: *Crassula ovata*, *Pachira aquatica*, and *Dracaena sanderiana* are all categorized as money tree plants. In the Latin naming

PROFILE:
Jeff Keil (he/him)
Owner/Head Grower Otto Keil

"I don't just keep my at-home plants happy, I also watch over thousands of plants at the nursery. Good care is a 24/7 job—plants don't take weekends off! Both at home and at the nursery, keeping plants happy means finding them the right light and temperature. I work hard to provide appropriate means of fertility for healthy growth. And at the nursery, you have to keep on top of the bugs!"

Mass production of plants is not easy! Some plants can be cloned using tissue cultures, but others must be painstakingly hand-propagated.

convention, the first word tells you the plant's genus (like a surname) and the second word tells you the plant's species (like a first name).[11] Common names for plants typically carry cultural meaning or allude to the plant's physical characteristics, whereas Latin names give you the plant's entire genealogy, informing growers about the plant's origin, its relatives, and what it needs to thrive.

Nurseries are living libraries of incredible botanical knowledge that has been passed down through generations of growers. Despite differences in varieties grown and methods of cultivation, all plant nurseries believe the uptick in growing houseplants does not signal a trend but a growing recognition that plants are essential to our lives.

Chapter Two: You

YOUR NATURE FOR NURTURE

Plant care is a practice rooted in the core of our beings, and our survival has always been dependent on the lush and expansive plant kingdom that surrounds us. Your curiosity about plants and desire to coexist with them is part of an ancient pull toward life. It is an evolutionary impulse to help sustain ecosystems that not only shelter, feed, and clothe you but also provide an endless source of joy and wonder. The unshakeable emotional attraction to the natural world—from forests and jungles to parks and, yes, your houseplants—is called biophilia, which in Greek means "the love of what is living."

Harvard biologist Edward O. Wilson and social ecologist Stephen R. Kellert popularized this term in their book, *The Biophilia Hypothesis*, arguing that humans are innately connected to nature and that our emotional attraction to the natural world underlies our biological composition.[1] Wilson found that our attraction to nature (biophilia) and fear of nature (biophobia) are both part of our genetic makeup and that we tend to fear natural

species, like spiders and snakes, more than our own dangerous human-made creations, like guns and cars. Interestingly, humans are more attracted to plants that are useful to us, like trees that are climbable or have large, shady canopies, and when all we can do is sit back and watch the world go by, humans would much rather view the ocean, vegetation, and flowers than stare at concrete.

Many science writers and theorists share the idea that humans have an evolving relationship with the natural world. In her book *Braiding Sweetgrass*, naturalist and professor Robin Wall Kimmerer writes that humans who did not grow up with Indigenous people's origin stories of Earth rarely understand our universal connection to the earth.[2] Kimmerer believes that once we learn about those Indigenous origin stories, we will feel a sense of urgency to actually do something about the current climate crisis. Our societies need to embrace biophilia in order to reverse our estrangement from the natural world and begin to see ourselves as collaborators with nature rather than masters of it.

Too often, especially in Western societies, people think of themselves as philosophically separate from the rest of nature and ordain themselves as conquerors over plants and fellow animals that inhabit Earth.[3] This behavior is linked to oppressive hierarchies within our own species that demand homogeneity in pursuit of power and that punish diverse expressions of humanity, like skin color, gender, sexuality, language, and culture, by labeling them closer to nature (i.e., primitive, savage, inferior) and, therefore, to be exploited.[4] Assuming superiority over other living beings creates a chasmic binary between humans and the environment, leaving people disconnected from the self and the role as a coinhabitant in nature.[5] But it does not have to be this way.

Light is food for plants, and the closer you place your plants to a window, the leafier they will be. The darkest shade outdoors is still many times brighter than a sunny window indoors.

YOU

Commonly known as a desert rose, this small plant with a big personality is planted in a handmade pot from Mexico City.

You are predisposed to connect with nature because you are a part of nature, and you have much more in common with plants than you might think. Not only do plants hold deep cultural and spiritual value for people but also, akin to humans, plants themselves have their own communities, families, and networks of shared knowledge.

Plant care is about more than keeping a plant alive—it is a way to facilitate curiosity about the vast and interdependent ecosystems that support you. Houseplants improve the quality of daily life by aiding in focus, improving air quality, and increasing feelings of calm. When you develop new rituals and plant-centered traditions that recognize your place within nature, you honor your responsibility to nurture your surroundings. The natural world is not something separate from your domestic life but surrounds your life every day. By sharing your home with plants, you invite daily reminders of the ecosystems you are a part of.

2.1
Becoming a Plant Parent

To bring a plant home is to embark on a spiritual partnership that transcends soil, water, and sunlight. Choosing your first plant is a bit like buying your first car or adopting your first pet—you would not take home the first one that catches your eye. Instead, you would research and make adjustments to ensure that your home will be a welcoming place. While sustaining plant life is not as involved as raising a child or adopting a pet, the care that it takes to grow a small sprout into a healthy herbage stems from the same root—nourishing love.

A close friend of mine came to Horti with a simple ask: "I want to get into plants, and I have a lot of light." After showing him a few options, we settled on a large leaf *Alocasia* × 'Calidora' (commonly known as elephant ears), for his sunny apartment. While his interest in plants was mainly dictated by the surge in houseplant-related content

Christopher Griffin (he/she/they)
Gardener/Content Creator

"I keep my green gurls happy by strutting into this viridescent plant-parent journey with a sense of curiosity, a desire to just learn. The more you let yourself wander into the vast world of horticulture, the better informed you'll be when it comes to caring for your green gurls. Learning about the history, origins, native habitats, and fun random facts about the green gurls I bring into my home has been quite helpful during this adventure. So whether it's casually reading articles online, picking up a gardening book, chatting with your plant frands, or connecting with your local plant shop, let yourself get lost in the greenery of it all, kween!" **@plantkween**

Becoming a plant parent is about understanding that your plant is a living thing and should be treated as such, not like a decorative object, as we often see in design magazines.

trending on social media, when he saw the first new leaf starting to rise and slowly unfurl, he realized this green object he brought home was actually full of life. This moment of mutual growth transformed his houseplant from a mere object in his apartment to a living creature with a name and needs. He called me in wild amazement and was so genuinely thrilled by this new development that he kept referring to the plant as his baby. And just like the friend in your circle who sends you too many photos of their newborn, he began updating me on every new leaf, his plant's rotation schedule, and whenever a leaf turned yellow by sending a sad emoji. Just like our own bodies, it is easy to take a plant's inner workings for granted and not bother understanding their complex biochemistry and behaviors

YOU

OPPOSITE
**If you don't get enough
light throughout your home,
shelving units are a great
way to accommodate
your large plant collection
around a window like Phoebe
Cheong's (@welcometothe
junglehome) Brooklyn
apartment.**

until something changes, like new growth or symptoms of sickness.

Plants are personal. Inviting them into your home means enmeshing their needs with your lifestyle, so it is important to ask yourself what your goals are when it comes to plant care. What do houseplants represent for you? Are you looking to improve your own wellness by caring for another being? Do you crave the presence of houseplants for a streamlined aesthetic?

Plants are not objects—they are more like housemates who do not pay rent but pull their weight in countless other ways. While some of your belongings may end up burdening you rather than comforting you, investing your time and resources into plants goes hand in hand with reduced consumption, limiting waste, and appreciating what you already have.

Plants are not cheap. Modern buying culture is designed around disposability, and, too often, houseplants are treated like the pet store goldfish of the plant kingdom: disposable tokens of sympathy or celebration that are quickly neglected, lucky to survive a few weeks in a new home. The notion that houseplants are disposable results from how plants are typically sold, with little information about their origins and their needs. Big-box garden centers stockpile greenery, and unsold plants are eventually sent to landfills, where they contribute to carbon emissions. The short-lived exchange of traditional plant buying reinforces common misconceptions that houseplants are finicky, hard to care for, and easier to replace than to maintain. But the truth is that plants can be long-term companions that will grow alongside you for years to come.

2.2
Your Lifestyle
and Home

The rituals of tending to houseplants will look a little different for everyone. Some forms of plant care involve steamy weekend showers, gentle daily misting, or meticulous pruning, while others might incorporate regular propagation and sharing cuttings with friends. You might talk to your plants over morning tea, meditate on their growth, or show them off to visitors and internet acquaintances. Whatever routines you develop with your plants, you will soon find that creating space for plant care does as much for your own well-being as it does for your plants. The time you spend feeding and grooming them doubles as moments for quiet reflection and a necessary pause from the outside world's endless demands. Houseplants offer an outlet for your need to nurture and remind you of the fundamentals of a happy life: clean water, fresh air, nourishing food, and strong roots.

Consider your environment from your new plant's perspective. Unlike pets, plants cannot vocalize their problems or relocate themselves to avoid harm. It is up to you to decide where your plant will thrive and to diligently monitor their happiness. Let us revisit some the basic conditions to consider and dig into them a little deeper.

Light

Light is the most important factor when it comes to your plant's happiness. What matters most is the direction your windows face, the climate you live in, and what your individual plant needs. Determining the direction your windows face—and what kind of light (direct or indirect) your home has—will save you time, money, and frustration when shopping for a houseplant. To help you navigate this new territory, we have curated a guide to common houseplants (see page 135), but if you are unsure about where to

LEFT
This six-year-old ZZ plant
receives bright direct
sunlight from the west for
a few hours every day.

RIGHT
If you are unsure about
the light needs of a plant,
placing it by a window
is always a safe bet.
However, be mindful of
drafty windows during
wintertime.

place your plant in terms of light, start by setting it in a sunny southernmost facing window and look for changes in coloration and growth over a few days. For sun-drenched homes, cacti and succulents make happy companions, but if your lighting is spotty or indirect, you may want to try a pothos (*Epipremnum aureum*).

Temperature

The temperature needs to be stable wherever your plant lives. If you are growing in a window, press your hand to the glass to feel the temperature before setting your plant down. Does the glass feel warm or chilly? Some plants tolerate one over the other better. For example, some orchids, like the Australian Pink Rock orchid (*Dendrobium kingianum*), do not mind a chill that dips down to 40°F (4°C), while other plants—like most tropical plants, succulents,

YOU

LEFT
Pots should always have drainage holes and saucers to allow water to drain freely and to provide air to the roots.

RIGHT
Some plants can benefit from extra care. Misting can help minimize the risk of overwatering.

and cacti—prefer to be in warmer spots. If you are growing under lights, consider the temperature of the bulb and the plant's proximity to it. You want your plant to be close enough to maximize the light coming from the lamp but far enough away to avoid getting burned by the heat.

Airflow

Some tropical plants benefit from airflow, especially if they are kept perpetually moist. This is most applicable in terrariums, which should never be sealed but remain open at the top to allow the plants inside to breathe. When spring arrives and you throw open up your windows, do not forget about your plants sitting on the sills. Temperatures in

Bryana Sortino (she/her)
COO, cofounder, Horti

"I would say it's all about placement! My apartment gets 360 degrees of light, so if I don't have a plant in the right spot, I'll lose it. With so much light, I frequently rotate and repot my plants. Rotating is important so that each side gets just as much light as the rest. With more than one hundred plants under my wing, I've figured out the exact timing for when a plant needs a repot. Developing this knowledge has been the first step in learning to parent, and it has definitely prepared me for being a mom to my puppy, Roo!"

both the spring and fall can unexpectedly fluctuate and leave your plant suddenly singed by heat or frostbitten by the cold.

Frequency

Some plants need you to check up on them. Air plants (*Tillandsia*) and maidenhair ferns (*Adiantum*) require a lot of attention and benefit from being pampered. If you enjoy regular spritzing and helicopter parenting, these are the plants for you. If you are a busy bee or forgetful by nature, it may help to set calendar reminders to check on your plants. If you are a frequent traveler, perhaps choose plants that can tolerate being left alone for extended periods of time.

Regardless of your plant species and home environment, the best practice is to check on your plants every few days, even if you do not water them. Be willing to get your hands dirty: stick your finger an inch or two into the soil to gauge how fast the soil is drying out. Inspect the plant's leaves and stems for signs of pests and sickness. And even if your plant has a clean bill of health, stroking its leaves stimulates receptors and encourages growth— much like petting a cat or dog. More than anything, enjoy your plant. After all, plants are more than just decorative objects, they are living creatures.

Pets and Plants

If plants are not the only creatures in your home, you will want to determine which ones are pet friendly and which are toxic. Most plants that are poisonous to animals are only harmful when ingested, so keeping your plants out of reach from curious nibbles and bites is key. That being said,

While some houseplants are toxic to pets upon consumption, most are harmless to the touch.

there are indeed plants that produce actual poisons and are highly toxic, such as *Euphorbia* and *Ficus*, so avoid bringing them into your home if you are a pet owner.

If your pet refuses to leave your houseplants alone, consider why they may be interested in the first place. Most vets agree that a curiosity in plants could be a sign that your pet is bored and needs more stimulation. If your pet and your plant are competing for windowsill space, consider placing your plants up high or in rooms that your pet uses less frequently, so they can both coexist in the same home peacefully.

YOU

Chapter Three: Your First Plant

A POTHOS GROWS IN JERSEY CITY

Finding a new houseplant is a special thrill, but physical appearances and first impressions aside, how do you really know if you and your houseplant will be the perfect match? Just like with any failed relationship, any unfortunate previous experience with plant care can affect if you view yourself as a plant person and whether you believe you are someone who can nurture life.

In 2010, I started my plant journey after moving into my first apartment in the US. My home was in a sparse four-story walk-up in Jersey City, New Jersey, where my partner and I lived with nothing more than a mattress and the two chairs that we had brought with us. Before investing in any furniture, I bought a large vining pothos that I spotted at a local grocery store. For some reason, I gravitated toward this plant, as if it chose me. Bringing my new plant into our home flooded my senses with waves of nostalgia. I felt the comfort of my childhood home in India, where I had grown up surrounded by my mother's houseplants. Before that moment, I had never thought of myself

as someone who cared about plants because my association with plants was tied exclusively to my mom's interest in gardening. Plants were my mom's thing, not mine. Within a few weeks of getting my first pothos, I started filling all of the empty corners of our apartment with plants, which brought a sense of calm and added beautiful textures to our tiny, empty space.

Houseplants created a sense of grounding as I found a new home in this country. While I definitely lost a few ferns in the process, I also learned a lot about my own temperament in caring for my plants. The pothos thrived despite my beginner missteps—as if the plant's resilience gave me a reason not to give up on planting, even after my failures. I kept the pothos at the entrance of our apartment directly under a skylight, and by the time I moved from that apartment the following year, its vines had grown almost two stories long. I decided to leave the plant behind because it was thriving where it was, and the new tenants gladly adopted my beautiful green creature. I took a small cutting from it to repot in my new house, and I can happily say that the same pothos is still with me. Since then, I have shared its love with many different people through propagating its cheerfully resilient vines.

Over the years, my business partner, Bryana, and I had grown accustomed to being sought out by our friends for planting advice since we both had almost one hundred plants in our respective apartments. We realized that, just like our friends, most people were buying plants for their aesthetic value without understanding how to care for them. We decided to curate a selection of plants that could withstand the learning curve of novice indoor gardeners, and Horti was born. It was important for us to build this as a subscription model, not just to secure recurring orders but also to guide our community through the stages of growth and slowly introduce them to new varieties over the course of months and years. We started with easy-to-care-for varieties and then slowly introduced other varieties each month.

I often wonder if had I not chosen an easy-to-care-for plant, like the pothos, how different my plant journey might be today. For first-time growers, finding a forgiving houseplant that bounces back from novice mistakes can keep your confidence from sagging, allowing you to learn as your plant grows.

3.1
The Five Best
First Plants

Pothos
Epipremnum aureum

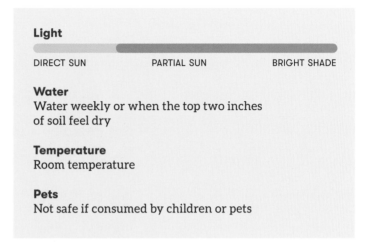

Light

DIRECT SUN   PARTIAL SUN   BRIGHT SHADE

Water
Water weekly or when the top two inches
of soil feel dry

Temperature
Room temperature

Pets
Not safe if consumed by children or pets

Pothos are opportunistic climbing vines that hail from
French Polynesia and are staples of the horticultural trade.
A descendent from a swamp-dwelling ancestor shared by
aroids, the pothos has retained its ability to be propagated
in water and has variable morphology—meaning that the
shape of the leaves and the plant will change throughout
its lifetime. Pothos are often grown hanging, but the telltale
mark of a climber is that the leaves turn upward at the tips
of the vines in the direction they want to grow when they
cascade. That means that while the plant will continue to
extend, it ultimately wants to grow upward. While there is
no harm in growing pothos in a hanging basket and letting
the vines cascade below, the pothos will not thicken and
grow into the next phase of its life cycle until it can attach
itself onto a medium for upward growth.

Snake Plant

Sansevieria trifasciata

Light

DIRECT SUN   PARTIAL SUN   BRIGHT SHADE

Water
These plants want to be watered after a dry rest. Once the soil is bone-dry, wait a day or two before watering.

Temperature
Keep at room temperature or warmer—it will grow faster when warmer. Tolerates and enjoys hot extremes.

Pets
Not safe if consumed by children or pets

Probably one of the most misunderstood genera is the *Sansevieria*. While recent genetic tests have consolidated the genus *Sansevieria* into *Dracaena*, we will refer to them separately for practical purposes. Snake plants are errone-ously sold as ultra-low-light plants, when in fact they just die very slowly in the absence of light. In the African scrublands, their native environment, they bathe in full direct sun or light shade in areas with sparse vegetation. In botanical gardens, snake plants are often grown in controlled temperatures, alongside cacti and succulents, and can grow up to twelve feet tall. Unlike plants whose leaves reach a finite size, the leaves of snake plants con-tinue to grow as long as conditions are stable and the tips of the leaves are not damaged. Snake plants can even bud with pleasantly fragrant chartreuse green flowers if given enough light. Treat the snake plant like you would treat a succulent: lots of sunshine, dry air, and very infrequent watering.

Baby Rubber Plant

Peperomia obtusifolia

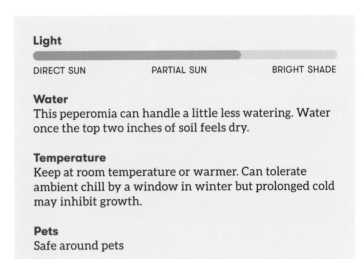

Light

DIRECT SUN PARTIAL SUN BRIGHT SHADE

Water
This peperomia can handle a little less watering. Water once the top two inches of soil feels dry.

Temperature
Keep at room temperature or warmer. Can tolerate ambient chill by a window in winter but prolonged cold may inhibit growth.

Pets
Safe around pets

Hailing from the neotropics, this plant is part succulent and easy to grow. The tall cream-colored spikes of the peperomia signal that the plant is flowering and that it is happy. Though it is not edible, the peperomia is related to the edible peppercorn vine and is pet friendly. This plant can be easily propagated by stem cuttings.

Spider Plant

Chlorophytum comosum

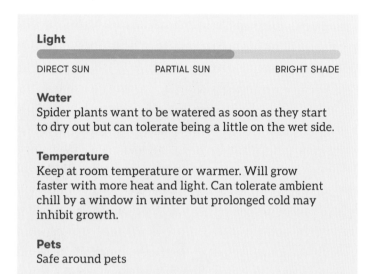

Light

DIRECT SUN   PARTIAL SUN   BRIGHT SHADE

Water
Spider plants want to be watered as soon as they start
to dry out but can tolerate being a little on the wet side.

Temperature
Keep at room temperature or warmer. Will grow
faster with more heat and light. Can tolerate ambient
chill by a window in winter but prolonged cold may
inhibit growth.

Pets
Safe around pets

The spider plant is well known in the houseplant world.
Hailing from East and Central Africa, this plant can live on
land or in water and is famous for its ability to grow long
strings of little plantlets from the mother plant. Spider
plants are fast growers, and in Florida, they are considered
to be noxious weeds that are routinely removed from
protected lands because they outgrow the local flora. Their
weedy traits will keep them alive and robust in many
home conditions.

ZZ Plant

Zamioculcas zamiifolia

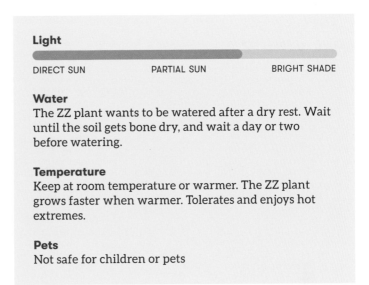

Light

DIRECT SUN   PARTIAL SUN   BRIGHT SHADE

Water
The ZZ plant wants to be watered after a dry rest. Wait until the soil gets bone dry, and wait a day or two before watering.

Temperature
Keep at room temperature or warmer. The ZZ plant grows faster when warmer. Tolerates and enjoys hot extremes.

Pets
Not safe for children or pets

Last but not least is the ZZ plant. This East African succulent grows in forest clearings or at the edges of subtropical forests, where it is exposed to direct sunlight and has been observed growing in gravel and soil. Like the snake plant, the ZZ plant is often sold as a "low light" plant, but the ideal care for this plant to grow large is direct sun in rich potting soil and watered after the soil completely dries out. As the ZZ plant grows, it tends to sprawl out instead of up, and you can remove leaves to prompt new growth.

3.2
You Got a Plant... Now What?

The exhilarating feeling that comes with bringing home a new plant can quickly turn into panic when you suddenly realize that you have no clue what to do with the tender green creature in your hands. In this section, I will dive into the three easy steps you need to take to successfully welcome your first plant (or second, or third) into your home: quarantine, repot, and rest.

Quarantine

By now, you know the ins and outs of quarantine all too well. When it comes to your newly acquired plant, a brief period of isolation is an important first step to ensure that you do not accidentally introduce unwanted bugs (which can often hide out in nursery plants) into your home environment. If this is your only houseplant, you do not have to worry as much about isolating a lone plant, but plant quarantine is a good rule of thumb to follow if you have an animal companion or a small child. If you already have other plants, this is an absolute must to properly integrate your new plant and keep potential spider mites, fungi, or other pests from infesting your healthy house-plant collection.

Place the plant in a room that does not have other plants in it, or if that is not an option, place the plant as far away from your other plants as possible for one week. In this initial quarantine, it is fine if the light is not ideal—this placement is temporary.

If you want to ensure that your plant is totally bug-free, I recommend spraying it with a natural insecticide solution. Simply dilute whatever natural soap you have on hand, such as Dr. Bronner's, by mixing one teaspoon of soap per quart of water in a spray bottle. Spritz your plant's leaves from all angles, wetting the top of the soil and the

Quarantine and treat all new houseplants in isolation for a few days in order to remove any pests and prevent them from infecting other houseplants.

outside of the nursery pot. Allow the solution to completely dry and repeat the process a second time if needed. After a few days, gently rinse the plant in the sink or shower (depending on its size) to remove any residual insecticide. Now you are ready to repot your plant.

Repot

Believe it or not, plants are often just as big underneath the soil as they are above it. Typically sold in temporary containers that are several sizes too small, any plant you bring home needs to be repotted into a larger container so that its roots can stretch out and grow. Look for a pot that is at least one to two inches wider in diameter than the one that your plant came in.

It is important to choose pots that accommodate the plants that live in them, with careful attention to function, style, and affordability. There are countless potting container options, which can make it difficult to choose the best one. I recommend using terra-cotta, which is made from natural clay rather than synthetic or petroleum-based materials. Terra-cotta pots have breathable yet sturdy walls that aerate the roots of the plant while providing added insulation, thanks to the heat they retain from the sun. Unlike flimsy plastic or dense concrete, clay mimics your plant's natural habitat and effectively slows down the transfer of heat between the internal soil and external environment, which helps to stabilize any sudden shift in temperature.

Once you have chosen your pot, you need to fill it with soil. Did you know that nearly 90 percent of plants thrive in regular potting mix? While there are special blends of soil out there (see page 101), the potting mix you need to keep most plants happy can be found at your local hardware store or garden center.

After quarantining your new plant, it is time to repot. This is an opportunity to select a stylized pot and give the plant more space to grow.

Rest

Being a plant parent is all about acknowledging the importance of adjustments, for yourself and your houseplants. Keep in mind that up until a week or so before arriving on your doorstep, your plant spent most of its life living in the perfectly controlled conditions of a greenhouse, with ample sunlight and humidity. Your home is a very different environment, and your plant will need time to adjust. This adjustment period will take anywhere between three to seven days and will typically manifest in your plant losing some of its leaves. When this happens, do not worry. Dropping a leaf or two is a natural part of your plant's adjustment process.

Leah Kirts (they/them)
Vegan queer writer

"My plants are happiest when I treat them like my friends. I visit with them, sit close enough to see what's going on with their soil and leaves, do what I can to make them comfortable, and make adjustments to keep them feeling cute. I would never force a meal or drink onto a friend, so I don't force my plants to glug down water without checking to see if it's what they need. I can't promise them an ideal climate or radiant sunlight every day, and I know they won't always be their leafiest, perkiest selves. Neither am I, to be honest. So I cheer on their summer growth spurts and wait out their winter slump. Doing so reminds me to treat myself and others with tender care."

Chapter Four: You and Your Plant

THE ELEMENTS OF PLANT CARE

Sharing your home with houseplants is a bit more involved than simply rearranging the furniture and identifying the sunniest corner in every room. Much like adopting a pet, plants require you to adjust your daily schedule and what you do for leisure. Nothing sparks joy like being surrounded by a lush, leafy jungle that you grew and groomed yourself. The trick to maintaining a healthy long-term relationship with your houseplants is to slow down, pay attention, and learn to observe how your plants respond to their new environment.

Plant care is a form of self-care, and the caretaking habits you form around your houseplants will inevitably be reflected in how you take care of yourself. Plants are daily reminders that you too are growing in ways that you might not be aware of. You will discover growth unexpectedly occurring in perky sprouts and unfurling leaves. Similar to your own evolution, plant growth is not easily measured from day to day but is best observed patiently over weeks, months, and years.

The elements of plant care are simple: light, water, and soil. Plants need these three things to create their own food and to thrive, but each component has its own quirks and complexities. In this section, I will explain why not all light is created equal, how your plants want to be watered (it is easier than you think), and the dirty truth about soil.

4.1
Plant Food

Light

There is an old horticultural adage, "The darkest shade outdoors is still brighter than direct sun indoors," which is an apt reminder that indoor environments restrict the intensity of sunlight that your plant is exposed to, no matter how sunny your home is. Plant growth is driven by sunlight, so more light ensures that plants will grow fuller and bushier. As there is less light indoors than there is outdoors, it is very difficult for your plants to be sunburned. What people mistakenly think of as sunburn on indoor plants is more likely to be a fungus. Sunburn causes the leaves of a houseplant to turn white or charcoal black, whereas fungus turns the edges of plant leaves yellow. The only time that sunburn is a serious threat is when a plant is moved outdoors without slowly acclimating it to the changes in their environment.

I am often asked which plants can survive in sunless basements or rooms that get little to no natural light, and I always disappoint people when I tell them that they really should not place a living plant in a room with insufficient light. As selling houseplants becomes increasingly profitable, it is easy for plant companies to lean into consumer demand and oversell certain varieties as "low-light plants." However, light is the most important factor that will determine the long-term survival of all plants.

OPPOSITE
Phoebe Cheong's cat, Pixel, at home, sitting in front of a sun-drenched plant shelf.

When plants are receiving enough light, they will produce robust new growth that can help them hold themselves up under their own weight.

Plants are brilliant energy gatherers. While they do not generate energy, plants collect light and trap energy in sugar molecules to both feed themselves and to create essential defense mechanisms for disease resistance against pathogens.[1] Plants restrict sugars from going into areas that are being actively attacked by fungi or bacteria to starve out the attackers and, instead, unleash an arsenal of chemical compounds made using the energy of the salvaged sugar molecules to defend itself. If your plant is not getting enough light, it will not be able to fight off rot or other diseases. Increasing your plant's access to light is often the best way to prevent fungal and bacterial infections.

Phoebe Cheong (she/her)
Photographer

"There's no secret formula to this! In order to keep my plants happy, I start by taking care of myself first. When I focus on my well-being, it puts me in the right mindset and place to care for others. In return, my plants are a constant reminder to take care of myself. Practicing plant care is also a way to become in tune with my mental health and stay grounded. That is why when I see my plants flourishing, it makes me feel happy and accomplished."

Did You Know: Plants Can Tell Day from Night

Plants measure daylight by periods of uninterrupted darkness using a protein called phytochrome.[2] During daylight hours, these proteins become active and trigger genetic changes and metabolic effects within the plant. At night, the inactive phytochrome essentially goes to sleep.[3] When darkness is interrupted, even for a second with a bright flash of light, the plant will think that it is daytime again.

Plant Placement in Windows

In the northern hemisphere, the sun rises in the east, swings to the south, and sets in the west. South-facing windows are ideal spots for plants because they can get soaked in sunlight all day, but the view your window overlooks can be just as important as the cardinal direction it faces. Wherever you place your plant, make sure that the view is not obstructed. If you have a south-facing window that faces a brick wall, for instance, it may as well be a northern window.

Bright, direct sunlight is the best for sun-loving plants like *Cattleya*, *Ficus*, succulents, and palms, which like to sit close to windows that have been warmed by the sun. Plant species that crave less direct light, like the genus *Calathea* and ferns, can happily sit in the shade of other plants, which helps to mimic a real forest. Just make sure that your plant is within three feet of a window to absorb adequate sunlight.

Since plants need sunlight to photosynthesize, they should be placed directly in front of windows, not to the

OPPOSITE
To understand how much light a plant is receiving, you have to envision what your plant "sees."

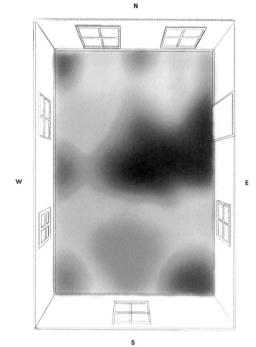

N

W   E

S

RIGHT
Light fades quickly indoors based upon distance from the sun, and your windows may let in less light than you think they do. In the northern hemisphere, the sun swings to the south midday, and thus southern windows with direct sun exposure are prized for indoor plant growing.

OPPOSITE
Artificial light is necessary when you do not have enough natural light from your windows to grow the houseplants that you want indoors.

side of them. Determine if your plant is in a good spot by sitting or kneeling down so that you can observe your plant's surroundings from its vantage point. Can you see out the window? Ideally, your plant should be placed in the center of the window or within the side of the window frame.

Artificial Lighting

All creatures active during the day benefit from good lighting. There are two elements of light to consider: quality and quantity. Quality is measured in wavelength, which corresponds to color. Quantity is measured in lumens and represents how many photons or light particles are coming from any given direction. Research shows that the direction of sunlight and length of exposure can make a substantial difference in the growth, fragrance, and, when applicable, taste of plants.[4]

If you are considering adding grow lights to your plant-care routine, there are different kinds of light, each with their own benefits. I recommend fluorescent or LED lights for your plants. While different colors of light have different effects on plants, it is best to replicate the sunlight that plants get in nature. Plants use the power of the sun as their fuel, and white light is the color closest to daylight. If you are considering using grow lights to supplement the quality of sunlight in your home, a spectrograph in nanometers may be handy (it will say if it is in nanometers on the package), which will tell you the actual wavelengths and lumens of the bulbs.

Photographers have developed an easy-to-use scale called the Color Rendering Index (CRI), which measures how close the wavelengths of artificial light mimic the sun. The CRI measures light on a scale of 0–100 with 100 being equal to direct sunbeams outdoors on a cloudless day.[5] Look for a grow light bulb that has a CRI rating of 95 or higher.

Water

A close friend of mine had a perfectly green and wavy bird's nest fern sitting next to his kitchen sink for almost two years. Being close to the faucet made it easy for him to remember to water his plant. He was so proud of his ability to keep this fern looking healthy that he brought more plants home. After two years of diligently caring for his plants, he decided to travel and asked a friend to water while he was away. When the friend came over to water the plants, she immediately discovered that the fern was plastic and the container had a pool of water sitting in it. I love this anecdote because prior to this hilarious discovery, my friend had developed genuine confidence in his ability to look after his fern, which then resulted in him bringing other (real) plants home to water and nurture.

Water is the source of life, so it is no wonder that water is foundational to plant care. Unlike my friend's plastic fern, living plants interact with water as if they are giant straws pulling in water to fill their plant cells. Plants take in the majority of their water through their roots, pulling the water up through their shoots, after which transpiration evaporates the water from the leaves out the stomata, also known as leaf pores.

While it sounds like a lot of work, plants do not contribute much effort into moving water themselves—the power of heat and light does most of the heavy lifting for them. The process is not a push up from the roots below but a pull from the leaves above to bring water upward. As sunlight heats up the plant, water evaporates, which prompts the plant to draw water up from the roots to rehydrate the cell walls of its stems and leaves.

Almost all plants absorb water through their roots, except for epiphytes, which have evolved special structures to absorb water from their leaves as well as their roots. Epiphytes, such as orchids, bromeliads, and other air plants, can absorb water through their leaves at varying rates, so routine spritzing with water is helpful for their growth. Other plants do not care if water ever touches their leaves; in fact, you should avoid spritzing nonepiphyte plants because wet leaves will inevitably attract pests and fungi.

Just like human cells, plant cells are mostly made of water. Shaped like boxes, plant cells have walls made of various types of cellulose that are semipliable but hold their shape. The size of a plant cell is largely dependent on the availability of water at the time of their initial growth. If your plant is growing and you miss a watering, its leaves may become misshapen.

Woody plants have a rigid structure that holds their shape in place, but tender green plants stay propped up by

When you are unsure of whether or not your plant needs water, dig into the soil a few inches using your finger. If the soil is dry, consider watering or misting.

cell walls that resemble water balloons stacked on top of each other. Plant cells are drained through transpiration, so if the cells are not rehydrated, they will lose turgor pressure, leaving the plant less stable and visibly droopy.

How to Water

Rain does not fall on the ground at exactly four o'clock every Thursday—rain falls when it needs to. Likewise, your watering regimen should follow suit. The soil is your guide and will tell you everything you need to know—but you have to get your hands dirty first.

Stick your finger into the soil a few inches below the surface, and if it feels dry, then your plant needs water. If the soil feels moist or wet, then your plant is fine. You will find that the soil dries out at different rates during different

If you are ever unsure of how much water to give your plant, run the plant under lukewarm (not cold!) water in the sink. Make sure the pot has a drainage hole. Always ensure that the soil gets fully saturated when watering.

times of the year, and even becomes water-repellent if it dries completely. If this occurs, slowly add water or pre-moisten the soil with a heavy spritzing before eventually returning to regular watering.[6]

My favorite way to water my small plants is to take them to the sink and let warm water run through their soil until it is fully saturated. Plants do not enjoy ice cold showers any more than most humans do. Cold tap water is a frigid 55°F (12°C), which is too chilly for most houseplants—use room temperature or warm water.

The goal is to saturate the soil of your plant each time you water it. You do not need to stop watering your plant at the first sign of water flowing through. Depending on how dry the soil is, let the water run through the soil for thirty to ninety seconds. One perk of watering your plant in the sink is that excess water can easily flow through the drainage holes in the pot and down the drain, mess free. Alternatively, if your plant is too large to fit in the sink, put it in the bathtub or shower. When watering in situ, let the saucer fill with no more than a half an inch of water, and allow your plant to sit for about a day before draining the saucer.

As previously mentioned, avoid splashing water onto the leaves of your plant. Wet leaves will leave your plant susceptible to fungal attacks. Unlike the outdoors, there is no indoor breeze to naturally dry the leaves off quickly, so take care to keep those leaves dry.

Water Quality

Water quality matters for everyone. No matter where your water comes from, minerals and other chemicals can leach into it. Some minerals, like calcium and magnesium, are beneficial to plants in small quantities, but too high of a quantity can be detrimental. Soft water is ideal for most houseplants because it has a low mineral content. While you want some minerals in your water, it is important to note that, over time, minerals will accumulate in the soil. Mineral accumulation is fine for outdoor plants but not for your houseplant. Hard water deposits too many minerals into your plant's soil at once and can cause the tips of its leaves to turn black—too much of a good thing can be fatal for your plant.

To safeguard your health as well as that of your plant, get your water tested to determine if you have hard water or if your water has other contaminants, like too much iron or acidity. Consider investing in a water purifier or setting a container in your backyard or on your fire escape to collect rainwater for watering.

A Brief Note on Humidity

———

Humidity is a measure of how much water is dissolved into the air. As humidity increases, transpiration or evaporation decreases, and vice versa. While most houseplants are not dramatically affected by humidity levels, all epiphytes and a few terrestrial plants are adversely sensitive to dry air. Orchids, bromeliads, air plants, *Calathea*, *Alocasia*, and some ferns benefit from being spritzed or placed near humidifiers in addition to being regularly watered. These plants live their best lives in terrariums or bathrooms where heat and moisture can easily be trapped. For most common houseplants, like pothos, *Monstera*, and fiddle leaf figs, humidity is not a priority. If your home has low humidity, you will need to water your plants a little more often to offset the drier air and increased water loss through transpiration.

As long as you give your plant attention, it will communicate what it needs . You cannot always achieve the optimum levels of humidity, frequency of watering, and intensity of sunlight because you are working with factors that cannot always be predicted or controlled. Understanding the basic elements of plant care will aid in working with, not fighting against, your plant's changing environmental factors.

Soil, Vitamins, and Fertilizer

———

When it comes to plant care, it is easy to focus on what is visible, but remember that 50 percent of your plant lives underground. Roots are just as important as shoots, but what is even more essential to growth is the material they both call home: the soil.

Most plants are planted in a soil mix that is conducive to their growth as young sprouts raised in a greenhouse—

OPPOSITE
Christopher Griffin's (@plantkween) beautiful indoor jungle getting some extra help from a humidifier to recreate the ideal environment for their tropical houseplants.

CHARCOAL

VERMICULITE

ORCHID
BARK MIX

COCONUT COIR

LAVA ROCKS

FERTILIZER

LECA

PERLITE

POTTING MIX

CALCITE

SPHAGNUM

not in a home. After you bring your plant home, repot it in a soil blend that promotes indoor cultivation. Most terrestrial plants will successfully grow in a general-purpose potting mix.

If you tend to overwater, or if your plant's soil takes a while to fully dry out, buy or create a mix that encourages better drainage. Depending on your plant, you can add sand, gravel, or lightweight expanded clay aggregate (LECA) to assist in healthy drainage. If your plant is an epiphyte or one that benefits from moist soil, consider putting small rocks, like pebbles or gravel, on top of the soil to help trap moisture. Give yourself some time to get acquainted with your plant before you dive into making your own mixes. For brand new plant parents, the best horticultural advice is to adjust your watering habits to your soil.

What's in Your Soil?

Regular potting mix is your default potting soil and is suitable for a majority of houseplants. Below is a breakdown of common additives to use to modify the soil to best suit your plant's needs.

PERLITE is a lightweight and popcorn-like white substance made from superheated volcanic glass. It can be used to aerate the soil by reducing the overall density in the pot. It is nonrenewable and extracted through mining but is otherwise safe to work with.

LAVA ROCKS increase drainage in soil for terrestrial plants, like lithophytes, or plants that grow on rocks. LECA, or lightweight expanded clay aggregate, is a kind of lava rock with a trendy acronym. You do not need lava rocks for drainage if you have a drainage hole.

VERMICULITE aids in water retention and supplies plants with trace minerals; however, the negative environmental impacts far outweigh the good. Vermiculite is extracted through mining, which releases asbestos into the air and harms the local ecosystem and the workers who extract it.

CALCITE is a natural pH buffer that helps maintain alkalinity in plants, such as *Alocasia* and *Paphiopedilum*, that are native to areas with limestone. Calcite is a natural source of calcium, an essential mineral for growth and disease resistance. Unpolished calcite is more effective than polished. Avoid using calcite on plants that do not need added calcium, or on plants that you do not know much about.

ORCHID BARK MIX is made with cedar and pine bark, along with charcoal, perlite, and sphagnum. Recommended for most orchids, *Alocasia*, *Anthurium*, *Hoya*, and other epiphytes.

SPHAGNUM is a type of dried moss that is often used as padding on plant mounts to help retain moisture. Dried sphagnum behaves a lot like wool—soaking up moisture when it is wet and repelling water when it is dry—so you must premoisten sphagnum to aid in absorption before watering your plants. Because sphagnum is slow growing, many sources harvest it from bogs, but some operations grow sphagnum sustainably. Pay attention to the source of your products.

COCONUT COIR, similar to sphagnum, is a little difficult to moisten when it dries out. I don't recommend using it by itself. At Horti, we use compressed coco that is mixed in with a ton of nutrients like mycorrhizae, kelp, worm castings, and magnesium.

CHARCOAL is only used for salt-sensitive plants, like orchids, some aroids, and the genus *Calathea*. Too much of it can increase your plant's pH or leach essential fertilizer salts, so use with caution.

Fertilizer is like a multivitamin for plants. The pot your plant lives in is its universe, so occasionally replenish the nutrients that the plant requires to thrive. Follow the instructions on the packaging for how and when to apply fertilizer, and be careful not to overfertilize, which can cause the tips of your plant's leaves to turn black as the plant pushes all of the excess salts it has absorbed to the tips of its leaves. Salt concentration in the tips of the leaves causes those leaf cells to dry out. If you think you have overdone it with a fertilizer, flush out your plant's soil for ten minutes with lukewarm running water. If the plant grows new leaves with black tips, remove it from its pot (terra-cotta and ceramic pots retain minerals and fertilizer salts), and repot your plant in fresh soil in a temporary plastic pot until it recovers.

4.2
Repotting

Houseplants quickly outgrow the flimsy plastic pots that most plants are sold in, and it is best practice to repot your plant within two weeks of bringing it home. Cramped roots

A spider plant can be kept root-bound for longer than most other plants.

will not immediately kill your plant, although they will eventually need more room to spread out and grow. Remember, plants are just as big underground as they are above the soil. If a plant that is forty-eight inches tall is stuffed into a six-inch container, it will not have enough roots to support the part of the plant above the soil.

An easy way to ensure your plant has a spacious home is to follow a rule of thirds. On average, the ideal ratio for a plant is for two-thirds of its height to be aboveground and one-third belowground. If the visible part of the plant exceeds this height ratio, it will need to be repotted or trimmed. Only trim the part of your plant that is aboveground—never trim the roots unless they are rotting, dead, or you are attempting bonsai, an artistic method of growing and training a plant.

Repotting is not just for new houseplants. Consider repotting your plants as part of your spring-cleaning, done at least once a year. Potted plants will stop growing new shoots if they have met their ideal root-to-shoot ratio, so you need to repot them in bigger containers to allow for underground root expansion and shoot growth above the soil.

When you prepare for repotting, especially if you have several houseplants, you want to first set out your *mise en place*—all of your repotting ingredients and equipment should be within reach. Set up somewhere you can make a mess without worry. If you do not have space outside, you can lay down an old sheet, towels, or a tarp. I recommend premoistening any soil or additives you plan to use in a large mixing bowl, keeping this mix damp while you work. If the pot has a drainage hole, there is no need for drainage rocks in the bottom of the pot, but if there isn't a drainage hole, use rocks and make sure they measure less than three-quarters of an inch in depth.

1. Place a small layer of moist soil in the bottom of the pot, and pack it tight—you cannot overpack the soil.
2. Remove the plant from its current container, and tug at the roots to loosen and untangle them from their original shape. (Do not cut the roots.) Try to remove about half of the old soil.
3. Place the plant in the center of the new pot, and add soil to all sides, packing tightly.
4. Cover the plant with soil up to one inch from the top of the pot, so that there is room for water to pool when you water the plant.
5. Always water your plant after repotting to bind your plant and its new soil together.

Choosing the Right Pot

At some point, every new plant parent will stand before heaving shelves of glossy ceramics, fluted concrete, and ridged terra-cotta in endless shapes, sizes, and patterns, scratching their heads with uncertainty. Pots should be curated much like the plants that live in them, with careful attention to function (drainage holes), style, and affordability, so that you can stay focused on the fun part—tending to your plants.

TERRA-COTTA is made of natural clay and is solid enough to anchor top-heavy plants without being cumbersome to relocate. This is my favorite pot because it is eco-friendly, and its neutral earthen hues blend seamlessly with nearly any style. Clay is like the raw denim of pottery, as exposure to sunlight gives it a unique patina as it ages. Terra-cotta pots are by far the most affordable, and the smooth, blank canvas makes it a perfect surface to add a little flair, like the painted flourish that distinguishes Horti pots from the rest.

GLAZED CERAMIC is great if you want a more stylish pot. The downside to this material is that it dries more slowly than terra-cotta, and many are without drainage holes.

PLASTIC dries the most slowly of all planters, which can be a good thing for plants that like to stay moist, like ferns. The disadvantage is that if you have a dark home, plastic might make the soil soggy, which can attract gnats or lead to fungus. Plastic is a petroleum-based material that takes hundreds of years to decompose, making it one of the least environmentally friendly pots.

METAL should never be used as a home for plants. Metal pots are made of cheap materials that easily corrode and poison plants over time. Even if metal is used as an outside pot saucer, there is still risk of metal contamination if water pools at the bottom of the pot. The best safeguard for your plants is to avoid metal pots altogether.

4.3
Moving Your Plant

Changes in Seasons and Environment

Plants do not uproot themselves and change locations in the wild—limit changing your plant's location, especially if it is growing well where it is. If conditions change, however, it can be beneficial to move your plant. For instance, if you have a pest infestation, quarantine the infected plant. If the window next to your plant becomes too cold or your heating system is on and your plant becomes too dry, consider relocating the plant to a more ideal environment. If your plant is not getting enough light, try to maximize the window space you have or invest in grow lights if your windowsill is full.

Bringing a Plant Outside

As all plants originate in the outdoors, they can often return without much of a problem. Most houseplants are from tropical or deserts climates, and it is best to bring them outdoors during the summer to give them the chance to grow tall and lush under intense sunlight in ways they cannot indoors. Keep an eye on the forecast if you move your plant outside—nighttime temperatures should be above 55°F (12°C) before you introduce your plant to the outdoors.

Plants need time to acclimate to the more intense sunlight outside, so place them under full shade for the first two weeks. Most tropical plants are happy to stay in the shade, but if you want to move them into partial sun after the two weeks, make sure they get no more than six hours of direct sunlight each day. For desert plants, like cacti, succulents, and other direct-sun houseplants, move them to partial shade for one week before moving into direct sunlight. Staggering exposure to intense direct sunlight allows plants to acclimate smoothly and thrive outside.

It is important to note that putting your plant outside does not mean that you can switch to autopilot in caring for your plant. Outside levels of heat and sunshine are more intense and change hourly, which means that your plant will dry out quickly and may need to be watered up to twice a day. It is a good idea to place a deep saucer under the pot, which acts as a reservoir to hold water your plant can use to quench its thirst on hot days. After the first rain, make sure the water is fully draining out of the pot so your plant does not drown. Do not, under any circumstances, put plants without drainage holes outdoors.

When fall arrives and temperatures drop below 55°F (12°C), it is time to bring your plant inside. But first, give it a

Melissa Male (she/her)
Content Creator

"I definitely find it beneficial to speak to my plants. When they blossom, I thank them for their beautiful flowers. When I introduce a new plant to my home, I give it a mini-intro to the plants around it and tell the plant I'm so happy it's here. It may sound crazy to some, but I truly believe it makes a difference!"

The easiest way to create more plants involves taking a clipping of a few nodes before immersing the node in water or soil to develop its roots.

haircut. Plants grow much faster outside than when indoors, and it is best to remove old growth and prepare them for winter. (This does not apply to desert plants like cacti and succulents.) Trim off any damaged leaves first, then remove between one-quarter to one-third of your plant's leaves—this will eliminate leaf dropping in the winter months when sunlight can be scarce.

After trimming, follow similar quarantining and treatment steps as when you first brought your plant home. Quarantine your incoming plant away from other house-plants for three to five days and spray with a natural insecticidal soap solution (see page 123). Spritz the plant from the tops of its leaves to the soil in its pot. Repeat spraying once or twice throughout the quarantine before giving your plant a good rinse and letting fully dry. This helps eliminate any outside insects that have made the pot their home before reintegrating your plant back inside.

As your plant grows and you become better acquainted, you will develop a greater understanding of its ideal conditions, making the change in seasons and moving from indoor to outdoor (and back in again) a seamless transition.

Puberty and Blooming

As your plant ages and grows bigger, it will hit plant puberty and attempt to reproduce. Signs of sexual development vary by plant type: tropical plants flower, ferns sporulate under their leaves, and succulents either flower or send off little clones of themselves called offsets, or pups. This stage of your plant's life is relatively easy—flowering is a natural part of a blooming plant, or angiosperm, life cycle. Orchids, aroids, pileas, cacti, succulents, begonias, geraniums, and most herbs eventually produce flowers in a range of colors, sizes, and aromas. Gymnosperms, like pines and conifers, do not flower but instead form cones, whereas ferns, mosses, and other ancient plants germinate using spores, which goes to show that plant sexuality is a spectrum of beautiful variance, just as it is in humans.[7]

Grooming and Pruning

Everyone needs grooming from time to time, and your plants are no exception. Monthly or bimonthly plant grooming does not mean you need to get out the clippers and start snipping away. It can be as simple as wiping the dust off your plant's leaves with a dry cloth, removing dead leaves, checking for insects, and assessing your plant's overall health.

Do not try "plant hacker" home remedies, like wiping mayonnaise on your plant or applying leaf shine to make leaves shinier (these methods will actually clog your plant's pores and suffocate them). If you notice any calcium buildup on your plant's leaves, wipe them with a paper towel using a 10 percent vinegar solution (9:1 water-to-vinegar ratio) or a 25 percent lemon juice solution (3:1 water-to-lemon-juice ratio).

Some plants, like this phalaenopsis orchid, cannot be propagated using cuttings but rather through seeds or tissue culture.

If your plant is getting leggy or growing in an unusual way, you can certainly trim it. Just keep in mind basic plant physiology when you are trimming: the petiole, or leaf stalk, only grows once, but stems will regrow. You have probably heard of pruning trees to prompt new growth, but did you know that a similar phenomenon happens in other plants? It is called apical dominance. The top bud expresses dominance through auxin hormones, which suppress the growth of other buds. When you pinch off the topmost bud of the plant, otherwise known as the apical bud, it forces the lateral buds to grow leaves and branches. Removing the apical bud allows other buds beneath it to grow.

4.5
Making More Plants

After regular grooming and pruning, you may wonder what to do with your plant cuttings. While composting is an option, consider propagating them, if you want to easily make more plants. Not all cuttings can be propagated, only stems with nodes, tubers, or rhizomes. Nodes usually have dormant buds with meristematic tissues (tissues that can become any part of the plant). When nodes are severed from the rest of the plant, a hormonal imbalance is caused in the buds that activates the meristematic tissues, causing them to form roots (or in some cases, shoots). In rare cases, you can propagate sections of leaves, like with the *Restrepia* genus, ZZ plants, and begonias.

Types of Propagation

There are two basic forms of propagation: sexual and asexual. The most common form of propagation is asexual, which is done by creating a clone of the parent plant from a cutting with viable shoots or by separating a runner plantlet that has already rooted in the soil. Sexual propagation occurs within the seeds inside of flowers and cones, or

indirectly through spores (as seen in ferns, mosses, and other primitive plants), making the offspring genetically different from the parent. Below are the most common types of propagation:

WATER PROPAGATION

Plant cuttings with flowing roots suspended in water may look beautiful, but water propagation is not ideal for terrestrial plants. While aquatic plants can thrive in a liquid home, water is devoid of important nutrients that a majority of plants need and can cause rot if not changed frequently enough. If using this method, cauterize the plant cutting with a flame before submerging it in water, and plant in soil once its roots grow a few inches long.

SOIL PROPAGATION

This is my favorite method because it best mimics what happens in nature. After cutting a plant, cauterize the wound or let dry for a few hours, then plant in soil. Some plants, like *Hoya* vines, need to be planted on their sides to maximize their ability to root.

AIR LAYERING

Propagating woody plants, like fiddle leaf figs, rubber figs, cinnamon trees, and citrus trees, requires a method called air layering. It is a complex process that requires cutting a diagonal notch above a plant node and tying a bag filled with sphagnum or moist soil around the wound in order to make the node root. Once roots begin to grow, the whole branch can be cut off and replanted.

SEEDS

This is the cleanest way to propagate, as seeds do not carry diseases or bugs. The downside of growing new plants from

OPPOSITE
Do not be afraid to clip or tear the plant babies away from their mother— they will live if the cutting includes enough roots or nodes to support new growth. If a plant has at least four to five leaves and an established root system, it is big enough to be propagated.

seed is that for most tropical houseplants, seeds take
months to germinate and can be finicky, unlike kitchen
herbs and vegetables, which grow and germinate with ease.
Consult the seed packet for directions and plant the seeds in
regular potting soil surrounded by warmth, light, and
adequate moisture.

Pests and Diseases

Nature is a complex web of life. Plants, like all other living
organisms, depend on a mutualistic relationship with
bacteria, microbes, and fungi to survive. Plants possess
defense mechanisms that help safeguard them from
disease infections, but even the strongest defense can be

4.6
**Plant
Sickness**

suppressed by harmful environmental factors. Pests play an integral role in keeping plant populations in check by infecting, infesting, and culling the weakest plants to make room for stronger ones to survive. Your living room, however, is drastically different from the forest floor or jungle canopy. Houseplants do not have to compete for survival, thanks to your tender loving care. But no matter how diligent of a plant parent you are, you will, at some point, have to deal with pests and learn how to nurse a sick plant back to health.

How Do I Know That I Have Pests?

If you see irregular growth or strange markings on your plant, they are most likely the results of the work of pests. Long before you see any bugs, you will notice their markings. For example, spider mites leave behind white dots where they chewed. You will find these marks underneath the leaves on the main rib and veins. Thrips—the worst of the indoor plant pests—on the other hand, make small sporadic scratches on the tops of leaves. If leaves appear to be deformed or curl in on themselves, inspect the underside of leaves for signs of nesting insects. If you do not catch an outbreak in the early stages, you will eventually be able to detect the insects themselves or trails of their waste.[8]

Adopt a preventative mindset by checking your plant for irregularities when watering. Scan the tops, bottoms, and joints of your plant's leaves and stems, giving your plant a quick wipe down with a dry cloth to potentially remove anything that may have blown in from outdoors.[9] Pests often enter your home through an open window or door, so if your plant lives on a windowsill, increase the frequency of your checkups in warmer months when you crave a fresh breeze. Another vector of plant pests is cut flowers, which can harbor mites and thrips. If you bring a bouquet of flowers home, consider displaying them in a room away from your plants.

Common Pests

SPRINGTAILS are tiny, light brown insects with oval heads and four segmented antennae that live in the soil. They love wet terra-cotta, and if you have a clay pot, you will likely come across them. The good news is that they are completely harmless. And as long as their population remains small (larger numbers can be an annoyance), they are beneficial because they consume harmful fungi. If they multiply and become a nuisance, reduce watering to let your plant (and its pot) dry out a bit and the springtails should disperse.[10]

FUNGUS GNATS look like tiny fruit flies and live most of their lives in the soil as grub worms before maturing into adults to breed.[11] Fungus gnats are harmless, but they signal that your soil is overly wet and is not drying out properly, likely due to overwatering, insufficient light, or too little heat.[12] The best way to get rid of fungus gnats is to eliminate the conditions that attract them: excess moisture. You can do this by allowing the soil to dry completely before watering and increasing the heat of your plant's environment to speed up the process. If the gnats persist even after you reduce watering, mix a thin layer of diatomaceous earth—a naturally occurring sedimentary rock that is ground into a soft powder—with the top inch of your plant's soil to rid your plant of them.[13]

APHIDS suck—literally. Living underneath the leaves of your plant, these insects tap into the vascular tissue of the plant and suction out the sweet sap inside, which is a vital fluid that carries nutrients to cells throughout the body of your

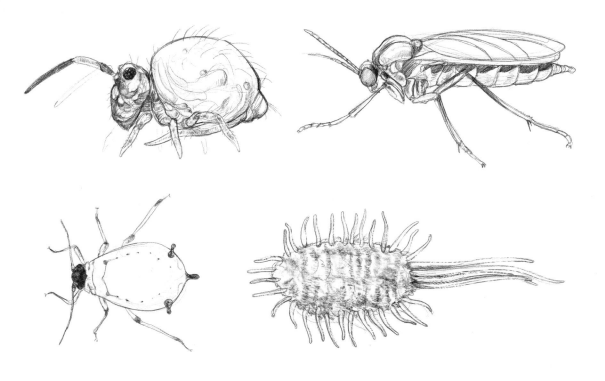

Clockwise from top left: springtail, fungus gnat, mealybug, aphid.

Plants and pests have evolved together for millennia. Some plants avoid being eaten by producing poisons or other secondary metabolites. Others produce physical deterrents like prickles, spines, and thorns to repel herbivores.

plant. You can spot aphids easily due to their tiny pear-shaped bodies, long antennae, and the sap droppings that they leave behind.[14] Like springtails, aphids are not a threat to your plant in small numbers, but their populations can quickly swell. The easiest way to get rid of aphids is to rinse them off of your plant with warm water. If aphids persist, spray your plant down with a natural insecticidal soap solution.[15]

MEALYBUGS are a lot like aphids in that they feed like vampires on the sweet sap inside of your plant and hide out under its leaves or among new growth.[16] Mealybugs are tiny pink insects with soft, oval-shaped bodies that are covered in a cottony wax. Their powdery white nests are water-repellant, and spraying them does little to disrupt their

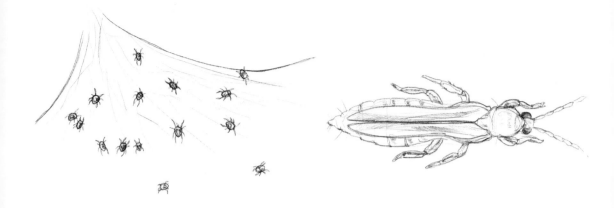

Left: spider mite
Right: thrip

Occasionally wiping your plant's leaves with a damp cloth is a good way to prevent pests from infesting your plant.

congregation. You can get rid of mealybugs by dipping a Q-tip in rubbing alcohol and wiping away all of the insects and their nests, then follow with an insecticidal spray. If one of your plants has mealybugs, it is likely that they all do, so check any nearby plants and treat them as necessary.

SCALES follow in the steps of aphids and mealybugs as sapsuckers.[17] Scales don't look like bugs at all but resemble small warts or scales (hence the name) and can often be mistaken as a part of your plant.[18] Known for their persistence, scales are hard to get rid of because their slim bodies can fit into hard-to-reach joints of your plant. The only way to kill them is to scrape off the adult scales and immediately spray your plant with insecticidal soap to kill the miniscule juveniles. Luckily, scales have a low reproductive rate compared to other plant pests and are manageable to eliminate but may require multiple treatments.

SPIDER MITES attack by slashing the plant's cells and drinking its fluids. Barely visible to the eye, spider mites use their webs to float in the wind and can infest all of your plants with their fine silky nests in one fell swoop. Unlike many other pests, spider mites are attracted to light and thrive in hot, arid conditions.[19] To rid your plant of spider mites, spray your plant with an insecticidal soap and thoroughly wipe down the stems and leaves. If you have more than one infested plant, be sure to wash your hands and change your clothes before handling other plants.

THRIPS are fellow slasher insects with fringed wings, red eyes, and narrow slender bodies and are pale green or yellow.[20] Thrips are known to curl leaves inward to protect themselves, sprinkling your plant with small dotted droppings.[21] Some species of thrips live in flowers, so be careful with any bouquets you bring home. Spinosad is an effective pesticide solution, but thrips can evolve to develop a resistance to pesticides.

Pests happen. They are an unavoidable part of growing plants. Thankfully, there are simple things you can do to help prevent pests and eradicate them quickly. Pay attention to your open windows, quarantine and treat your plants when they rotate back indoors in autumn, and use precaution every time you assimilate a new plant into your collection.

How to Keep Your Plants Pest Free

1. Treat and quarantine every new plant you bring home. Bugs are everywhere, so spray new plants with insecticide during their quarantine.
2. Use the correct pesticide. Different pests require different treatment—make sure that whichever pesticide you use is right for the bugs you are trying to eliminate (and safe for your plant)!
3. Repeat treatment until all pests are gone. Diligence pays off in your battle against bugs. Even if a new plant that does

not appear to have any pests, treat it within the first few weeks as a precaution.

4. Follow label instructions to the letter. Any antipest product you choose will have detailed instructions on the label. This is your gospel. Read the instructions, and do not deviate from them.

Types of Pesticides

The word *pesticide* might sound scary, but it is not necessarily synonymous with harmful chemicals. The pesticides that I suggest are approved for home use and often composed of natural ingredients.

INSECTICIDAL SOAP is an effective and nontoxic insect treatment, thanks to its potassium salts of fatty acids. Always dilute the soap in soft water (hard water will render it inactive), and never combine treatment with horticulture oil, as the soap will emulsify the oil, rendering it ineffective against insects.[22]

HORTICULTURE OIL is completely odor-free and an effective pesticide. As a petroleum product, it is not the most environmentally friendly, but it is toxic enough to bugs to kill them (while nontoxic to humans).

WETTABLE SULFUR works hard against mites and doubles as a fungicide. Sulfur is a plant nutrient and can help strengthen your plant's immunity to fungi. I recommend this method for more experienced plant parents. This is more commonly used as an outdoor pesticide, so be mindful when applying it to houseplants because without naturally occurring rainfall to wash the sulfur away, it can acidify the soil or burn your plant if it is used too frequently or in great quantities.

SPINOSAD-BASED INSECTICIDE is great for spider mites and thrips, and most pests in general. It is somewhat toxic to humans, so carefully follow the safety instructions on the label and avoid direct contact with skin. This should only be used after other treatment options have been exhausted.

INSECT GROWTH REGULATORS (IGRS) are preventative treatments. Rather than killing bugs, they prevent bugs from growing and reproducing. Most IGRs are food safe and can be used as an additive to other insect treatments. However, always read the labels carefully before mixing with another treatment.

NEEM OIL is often touted as a safe pesticide, and the strongest thing about it is the truly noxious smell. Neem oil is so safe and nontoxic, it is practically useless at getting rid of bugs, and I do not recommend it as an effective treatment. While the physical properties of oil can clog the breathing pores of an insect and block its airway, leading to asphyxiation, you have to douse the insect directly, which can be difficult to do if the pest moves quickly or if it is too small to be detected by the human eye.

Some plants cannot be saved. It is hard to accept it when a sick plant is too far gone, but as a rule, if a plant is more than 50 percent infested and costs less than fifty dollars, then it is not worth saving. Keeping a dying plant is like clinging to a bad relationship. Some of them just cannot be nursed back to health. Check with your local city composting and yard-waste programs to find out how to discard your plant in a safe and environmentally friendly way.

Dave Giglio (he/him)
Nutritionist

"I feed my plants leftover loose-leaf tea leaves because they are a good source of nitrogen. I water my indoor plants every Sunday during the quiet hours of the morning when my partner is still sleeping. I'll play Lana Del Rey for my plants (and dog) when I'm away from the house. This, along with Southern California sunshine, keeps them happy."

How Do I Know If My Plant Has Fungi or Bacteria?

———

Bacterial and fungal lesions will cause yellow discoloration that begins around the outer edges of the leaves and moves inward, leaving a path of dead tissue in their wake. Evidence of lesions on plants is often misunderstood as a sign of low humidity, which causes the edges of plant leaves to droop and turn brown. It is helpful to know the difference between these two common issues and to recognize infection at the first signs of discoloration.

To prevent fungal and bacterial infections, keep the leaves of your plant dry and do not spritz with water. Remove leaves that are more than 25 to 50 percent damaged, and thoroughly spray your plant with a fungicide, ideally with copper as the active ingredient. The copper is usually in the form of an emulsion or salt, depending on the formula, and works by arresting cellular processes in fungi. For orchids, mosses, and some ferns, use Physan 20, as these plants are sensitive to copper poisoning. Always follow the instructions on the label of any treatment you use.

How Do I Know If the Problem Is Abiotic?

———

Abiotic issues are caused by environmental factors that can lead to physical ailments in plants. Abiotic factors include having too much or too little water, light, minerals, or heat. For example, you might accidentally forget to move your plant off of the radiator at the end of summer and, when the heat kicks on, find your plant cooked to a crisp. Abiotic factors are the easiest problems to fix, and the damage is easy to spot.

Quick Guide to Signs of Abiotic Damage

ISSUE: TOO MUCH WATER

SIGNS: Leaves develop a splotchy pale yellow before turning school-bus yellow and dropping. Soil becomes soggy and tiny gnats appear.

SOLUTION: Check the soil before watering, and increase your plant's exposure to light and heat by moving it closer to an uncovered window or a warm radiator to help dry out the soil. Never put your plant directly onto the radiator, otherwise you will cook your plant.

ISSUE: TOO COLD

SIGNS: Your plant stops growing, droops slightly, and exhibits similar symptoms to an overwatered plant.

SOLUTION: Increase and maintain a stable room temperature, ensuring that nearby windows are well sealed. Reduce draftiness by sealing your windows with weather stripping or shrink film during the winter.

ISSUE: TOO LITTLE WATER

SIGNS: Leaves droop, curl, and turn brown around the edges. Your plant drops its leaves, starting from the bottom. For plants that cannot droop, such as cacti and succulents, they will begin to shrivel and wrinkle. The wrinkle lines grow deeper as your plant becomes thirstier.

SOLUTION: Soak your plant, and increase the frequency of waterings.

ISSUE: TOO HOT

SIGNS: Your plant dries out faster than usual and shows symptoms that are similar to an underwatered plant, but it also resembles a sautéed vegetable.

SOLUTION: Relocate your plant to a cooler room with stable conditions.

ISSUE: TOO MUCH LIGHT

SIGNS: White lesions appear on leaves that are most exposed to the sunlight.

SOLUTIONS: Decrease your plant's exposure to sunlight by moving it a few feet away from the window or by hanging a sheer curtain to filter out some of the sunlight. Indoor sunburns are rare in northern regions and higher latitudes. For example, a plant in a south-facing window in Southern California would burn, whereas a plant in a southern window in New York City would not. This is due to the angle of the sun in relation to Earth.

ISSUE: TOO LITTLE LIGHT

SIGNS: Your plant turns pale green and drops its leaves.

SOLUTION: Increase exposure to sunlight by moving your plant closer to an uncovered window or by using a grow lamp if sunlight is in short supply.

ISSUE: TOO LOW HUMIDITY

SIGNS: The tips of existing and new leaves appear to be burned, turning brown evenly around the edges. Plants that depend on humidity, like jewel orchids, will stop growing altogether.

SOLUTION: Use a humidifier to increase the water vapor in the air, or move your plant to a more humid environment, such as your bathroom, provided that your plant can get adequate sunlight.

ISSUE: TOO FEW NUTRIENTS

SIGNS: Plant growth is stunted, and the lower leaves appear to be bruised, yellow, and drooping. For houseplants, this is caused by nitrogen, phosphorus, and potassium deficiencies.

SOLUTIONS: Use a fertilizer to give your plant a boost of nutrients. Carefully follow the instructions on your fertilizer of choice. I recommend DynaGro or Good Dirt Plant Food. Always go with liquid fertilizers, which are more effective than pellet fertilizers or slow-release fertilizer spikes.

ISSUE: TOO SALTY

SIGNS: Leaf tips turn jet black from burns caused by excess fertilizer, salt, or hard water. Your plant may show wrinkles or signs of dehydration, and you may see salt accumulate on the pot.

SOLUTIONS: Repot your plant in new soil, and switch from unfiltered tap water to purified or softened water.

4.7
Soil to Soil

The world can be a harsh place, but plant parents foster a happier and healthier planet. You have probably been taught that your growth as a human is a constant pull between the oppositional forces of nature and nurture, when, in fact, nature is as much a part of our growth as we are part of its growth. Plants nurture human existence and help create and clean the air we breathe. Humans, in turn, help plants thrive alongside us in our home habitats—we water them in drought, shield them in flood, and guard them from prey.

Preparation is key to successful plant parenting. You should be prepared for the responsibility of sustaining life, the joy of watching your plant grow, and mourning the loss of your plant when it inevitably dies. No matter what your level of skill or your innate greenness of thumb, you will eventually say goodbye to one or many plants as part of nature's process. The death of your houseplant is not always a failure on your part—for some plants, it is just their time. In death, your plant has another purpose as organic matter in the afterlife, a necessary material found in rich, healthy soil.

Plants begin their lives as seedlings, pushing up from under the soil and spreading out across Earth until one day, just like humans, they bend low and shrink back down to where they came from, becoming an enriching part of the soil in their decomposition. After their fibrous matter is devoid of water and living cells, plants continue to contribute to the complexity of Earth's ecosystems. Being able to see beauty in this cycle of life and death is all about shifting your perspective.

Dead plants do more than contribute to the soil. Consider tea, the most widely consumed (and one of the oldest) beverages in the world. Arguably nature's most soothing elixir, tea is essentially made from dead leaves.

The very act of steeping and sipping tea requires us to slow down and inspires contemplation—both meditative habits that carry over into effective plant care.

We all have (or will one day have) prematurely lost our share of houseplants to beginner's neglect or obsessive smothering. The helplessness of watching a plant's health fade, or the tragedy of forgetting about it altogether, often discourages future attempts at plant care. Losing confidence in your ability to care for plants is a feeling of defeat that can last for weeks, months, even years. The problem in coping with the death of a houseplant is that it is not always obvious what to do next. There is no clear protocol and no easy burial. Grief can get swallowed up in the embarrassment that you were not able to sustain life, but it is important to reconcile your plant's death to move beyond it.

The best method for discarding your deceased houseplant is to compost it. Composting is a thoughtful way to commemorate the passage of plants from leafy and living to nutritious soil in the great cycle of life. Make sure that your plant did not have a fungus or disease that could be passed into the soil before depositing the plant into your neighborhood brown bins, community-garden compost shed, or a compost drop-off site at your local farmers' market, where you can often receive fresh compost in exchange. Donating your dead houseplant to become communal organic matter is as useful as it is therapeutic. You can liken the act of composting your plant to a public burial that will add important nutrients back to the soil, improving the quality of all life planted in it, from street trees and community-garden vegetables to your own future houseplants.

A GUIDE TO COMMON HOUSEPLANTS

Over the last few years, Horti has shipped nearly forty different varieties of houseplants to our subscribers. While our community learns how to raise their new plants, we have been on a parallel journey of learning how to grow our plant business and figuring out which plants make ideal companions for different lifestyles and environments. This guide covers some of the hardiest plants that we ship—and that I personally love! Just remember, plants crave their native environments, and understanding where a plant comes from and creating an approximation of it goes a long way toward keeping it healthy and beautiful.

NOTE: I have tried to be as botanically accurate as possible. Because common names like "money tree" often refer to more than one plant, I have included the Latin names of plants in order to be very specific as to which exact plant is being referred to. You will generally see the full Latin name, in italics with the genus capitalized, like Ficus elastica. Because there may be many species of a particular genus that can be cared for or treated the same way, I also may refer to them with the spp. epithet; e.g. Begonia spp. referring to all Begonia species (spp. = all species; sp. = one unknown species). Some plants are referred to by their family name here, rather than their genus or species names. This is because the plants of this group are so similar, we can refer to the whole family in terms of care. All plant families end in the suffix -aceae.

Staghorn Fern

Platycerium spp.

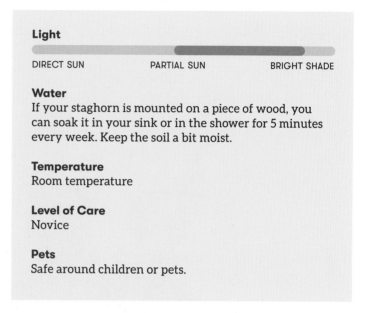

Light

DIRECT SUN   PARTIAL SUN   BRIGHT SHADE

Water
If your staghorn is mounted on a piece of wood, you can soak it in your sink or in the shower for 5 minutes every week. Keep the soil a bit moist.

Temperature
Room temperature

Level of Care
Novice

Pets
Safe around children or pets.

These ferns come from Australia and grow on trees in subtropical and tropical habitats. Unlike other ferns, staghorns have a tolerance for drying out because they have adapted to the hot Australian sun. Despite this adjustment, staghorns enjoy getting watered soon after they dry out. While they absorb the majority of water through their roots, the fuzz on their leaves is modified trichomes, or tiny hairs, that help ferns absorb water. Most staghorns are grown mounted on a wooden board with sphagnum. To water, remove the hanging board from the wall and place in a warm shower until it is fully soaked. Let it drip dry in the bathtub before hanging back on the wall.

Crispy Fern

Asplenium nidus

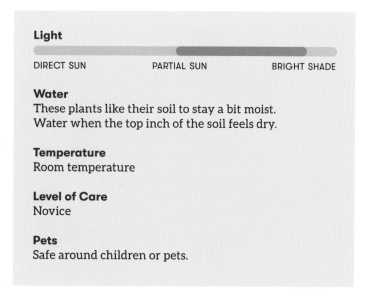

Light

DIRECT SUN   PARTIAL SUN   BRIGHT SHADE

Water
These plants like their soil to stay a bit moist.
Water when the top inch of the soil feels dry.

Temperature
Room temperature

Level of Care
Novice

Pets
Safe around children or pets.

Hailing from Southeast Asia, Australia, and the Pacific
Islands, these ferns grow in tropical climates, either epiphyt-
ically (on other plants) or terrestrially (in the soil). Before the
Permian–Triassic extinction, ferns were sun-loving plants.
However, due to the extinction event, which clouded the
skies with volcanic ash, the fern adapted to its new cloudy
environment through a horizontal gene transfer from a
low-light moss, helping it survive in the tropical understory
where the ferns can still be found today. Horizontal gene
transfer is a naturally occurring process of genetic modifica-
tion. In nature, plants swap DNA with bacteria and fungi,
and the bacteria and fungi swap those genes with other
plants. In this case, a fungus or bacteria took the understory-
survival genes and transferred them into an ancient fern
ancestor. Ferns are ancient, with a primitive vascular system
that allows them to grow only in moist environments, so
keep the soil damp.

Chinese Money Plant

Pilea peperomioides

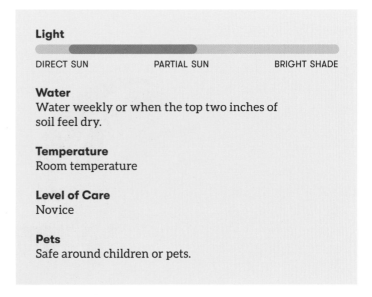

Light

DIRECT SUN   PARTIAL SUN   BRIGHT SHADE

Water
Water weekly or when the top two inches of soil feel dry.

Temperature
Room temperature

Level of Care
Novice

Pets
Safe around children or pets.

This plant hails from the Yunnan province at the foothills of the Himalayas in China, where it was collected by Western missionaries, lost in cultivation, and then rediscovered when a plant collector approached a botanist to identify the plant that had ended up in his private collection. While it is a part of the stinging nettle family Urticaceae, the plant lacks nettles and will not harm your pets. At remote monasteries in the foothills of the Himalayan mountains, this plant is used as a border, lining high-traffic dirt roads, thanks to its aggressive growth patterns. The more light it gets, the bigger it grows. Water when its soil fully dries out.

Warneckii

Dracaena deremensis

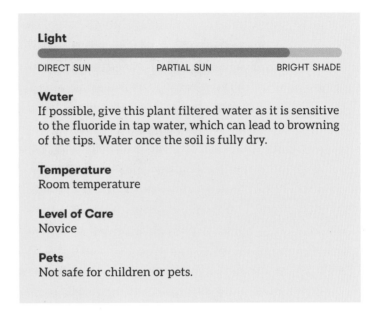

Light

DIRECT SUN　　　　PARTIAL SUN　　　　BRIGHT SHADE

Water
If possible, give this plant filtered water as it is sensitive to the fluoride in tap water, which can lead to browning of the tips. Water once the soil is fully dry.

Temperature
Room temperature

Level of Care
Novice

Pets
Not safe for children or pets.

Dracaena plants are pantropical, with their most diverse communities growing in Africa and the Arabian Peninsula. While most are grown ornamentally, one species, *Dracaena cinnabari*, can be used as a dye and is cultivated for its red sap, the only source for true dragon's blood incense. Most plants in the genus *Dracaena* are mistakenly sold as low-light plants, but they actually crave plenty of direct sunlight.

Begonia

Begoniaceae

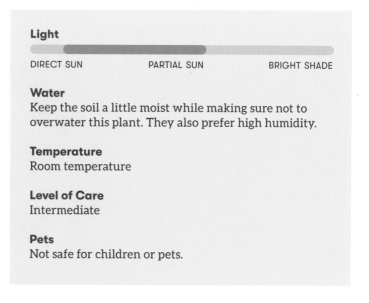

Light

DIRECT SUN · · · · · · PARTIAL SUN · · · · · · BRIGHT SHADE

Water
Keep the soil a little moist while making sure not to overwater this plant. They also prefer high humidity.

Temperature
Room temperature

Level of Care
Intermediate

Pets
Not safe for children or pets.

There are more than three hundred species of the Begoniaceae family. The pantropical plants are grown as much for their foliage as they are for their flowers. You can propagate begonias from any part of the plant, even a piece of a leaf. These plants like to dry out between waterings but cannot stay dry for prolonged periods. Any plant that is fleshy like a begonia has stricter watering requirements but will grow and recover quickly. Begonias are some of the fastest growers in the houseplant world and are great plants to learn with. They do best in east, south, or west-facing windows where they will receive direct sunlight. If you like to dote on plants, you will be successful with begonias, but if you are more neglectful, opt for succulents instead.

Satin Pothos

Scindapsus pictus

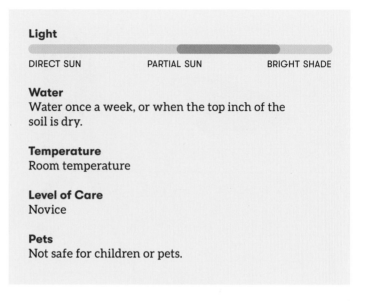

Light

DIRECT SUN · PARTIAL SUN · BRIGHT SHADE

Water
Water once a week, or when the top inch of the soil is dry.

Temperature
Room temperature

Level of Care
Novice

Pets
Not safe for children or pets.

The *Scindapsus* is a rarity in the plant world because of its "shingling" growth pattern, causing its leaves to press flat against whatever surface they are growing on. I recommend training them to grow up flat posts so that you can see their true form. Like other aroids, they are descended from a swamp-dwelling ancestor, so they retain the ability to propagate in water. These plants love humidity, so spritz them often to keep the soil moist.

Alocasia
Alocasia spp.

Light

DIRECT SUN   PARTIAL SUN   BRIGHT SHADE

Water
Alocasias like high humidity with a little break between waterings. Let the top two inches of soil dry before watering.

Temperature
Room temperature

Level of Care
Intermediate

Pets
Not safe for children or pets.

Alocasia are plants that have adapted to low-water environments by becoming dormant if they do not get enough water. They can be slowly brought back out of dormancy with a decrease in heat for a few months, followed by an increase in light and water. These plants are prone to mites, so take preventative measures (see page 120). Like other aroids, this plant has variable morphology throughout its lifetime, which means that the shape of the leaves and the plant will continue to change during its life. As a houseplant, the spade-shaped juvenile leaves will change shape when the plant reaches maturity (with size, age, and some direct sun).

Common Palm
Arecaceae

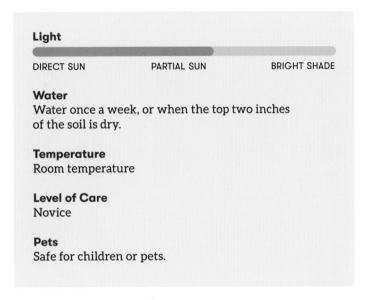

Light

DIRECT SUN　　　　　　PARTIAL SUN　　　　　　BRIGHT SHADE

Water
Water once a week, or when the top two inches
of the soil is dry.

Temperature
Room temperature

Level of Care
Novice

Pets
Safe for children or pets.

Palms belong to the Arecaceae family and need a lot of light
to thrive indoors. Many species are pan-tropical and are so
successful at long-distance seed dispersal that their seeds
have been found on some of the world's most remote
islands. While the most commonly grown palm houseplant
is the parlor palm (*Chamaedorea elegans*), other palms like
the coconut tree (*Cocos nucifera*) can be grown indoors. You
will need full floor-to-ceiling windows that receive direct
sunlight for a few hours in order for a palm to successfully
grow indoors.

Cow's Horn

Euphorbia grandicornis

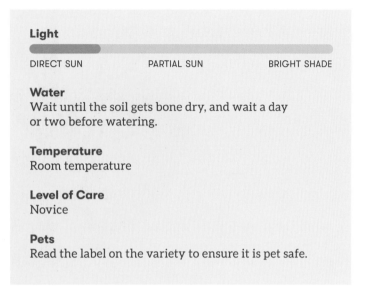

Light

DIRECT SUN   PARTIAL SUN   BRIGHT SHADE

Water
Wait until the soil gets bone dry, and wait a day
or two before watering.

Temperature
Room temperature

Level of Care
Novice

Pets
Read the label on the variety to ensure it is pet safe.

While all cacti are succulents, not all succulents are cacti. Many succulents are defined by their moisture-storing capacity and ability to sustain periodic drought, and some, like cow's horn, are just as hardy as a cactus. The plants of the genus *Euphorbia* hail from the Old World, largely Africa, and cacti come from the New World, the Americas, yet they have both evolved to adapt to similarly arid environments. Euphorbs bleed a latex-like sap that is toxic and irritant, whereas cactus sap is clear and mostly non-toxic. The flowers of a *Euphorbia* plant are unisexual— meaning each flower has only a male part (stamen) or a female part (pistil). There are over two thousand species of *Euphorbia*, and it is estimated that just under half of them are succulents with very similar care to that of the cow's horn: plant in well-drained soil, blast with as much direct sun as possible, and water thoroughly once the soil is bone-dry.

Zebra cactus

Haworthia attenuata

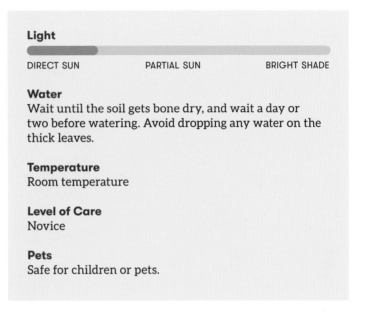

Light

DIRECT SUN   PARTIAL SUN   BRIGHT SHADE

Water
Wait until the soil gets bone dry, and wait a day or two before watering. Avoid dropping any water on the thick leaves.

Temperature
Room temperature

Level of Care
Novice

Pets
Safe for children or pets.

This plant produces an array of slow-growing, easy-to-care-for succulents. If you are looking for a low-maintenance plant, this is the one for you. Many *Haworthia* plants hail from the scrublands of Southern Africa, where strong winds and frequent dust storms caused them to evolve thick column-shaped leaves and translucent epidermal tissues, otherwise known as "leaf windows," which allow the plants to photosynthesize underground. Like any succulent, this plant thrives in direct sun. When you water, saturate its soil before letting it drain, and only add water once the soil is bone-dry again.

Rubber tree

Ficus elastica

Light

DIRECT SUN   PARTIAL SUN   BRIGHT SHADE

Water
Water once a week, or when the top two inches of the soil is dry.

Temperature
Room temperature

Level of Care
Novice

Pets
Not safe for children or pets.

Ficus is a pantropical genus, and the sap it produces was once cultivated to create rubber, hence its colloquial name, rubber tree. When the rubber tree grows to three feet tall, it throws off adventitious roots—roots that grow from non-root tissue—grabbing onto whatever is nearby. Unfortunately, these roots sometimes choke neighboring trees to death, earning it the nickname "strangler fig." They will grow as many leaves as light allows and like to be watered after their soil is completely dry.

Monstera

Monstera spp.

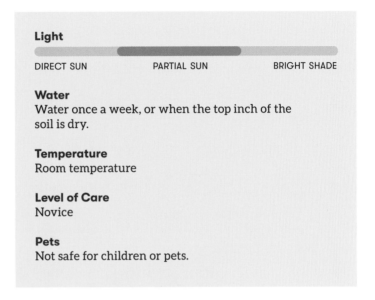

Light

DIRECT SUN · PARTIAL SUN · BRIGHT SHADE

Water
Water once a week, or when the top inch of the soil is dry.

Temperature
Room temperature

Level of Care
Novice

Pets
Not safe for children or pets.

These plants are neotropical vining aroids that, as descendants from a swamp-dwelling ancestor, retain their ability to be propagated in water. *Monstera* develops more leaf holes as it ages and grows. Do not cut off aboveground roots but train them to grow into the soil. *Monstera* will grow faster with more light, so try to provide it with as much direct sunlight indoors as you can if you want it to grow large. While this plant is inclined to grow horizontally, it can be trained to grow upward around a moss pole.

Philodendron hybrid

Philodendron spp.

Light

DIRECT SUN   PARTIAL SUN   BRIGHT SHADE

Water
Water once a week, or when the top two inches
of the soil is dry.

Temperature
Room temperature

Level of Care
Novice

Pets
Not safe for children or pets.

Philodendron is a large genus of plants with many morpho-
logical differences and whose leaves and overall shape
change throughout its lifetime. These plants are quite
resilient. In nature, they are one of the most adaptive
genera within the family Araceae, and inhabit many
environments. Most *Philodendron* species are climbers, and
their leaves will change shape as a result of their age and
environment. Like other aroids, these plants descend from
a swamp-dwelling ancestor and can be propagated in water.
While you can hang a vining *Philodendron* in a basket, this
plant prefers to grow upward. Care for these plants like any
other understory tropical plant: place in a window with
direct sunlight, water when the soil becomes dry, and keep
the temperature stable.

A GUIDE TO COMMON HOUSEPLANTS

Wax Plant

Hoya spp.

Light

DIRECT SUN   PARTIAL SUN   BRIGHT SHADE

Water
Water once the soil is fully dry.

Temperature
Room temperature

Level of Care
Novice

Pets
Not safe for children or pets.

The *Hoya* plant hails from a large region of Asia, and may also be found in Australia. Prized as houseplants for their semisucculent nature—and easy growth and flowering patterns—many *Hoya* plants grow as tropical vines. When it blooms, this vine can produce sweet, edible droplets of sap along its extrafloral nectaries. While they are generally safe to have around pets, cats and dogs may become sick if its leaves are ingested.

Peacock Plant
Calathea makoyana

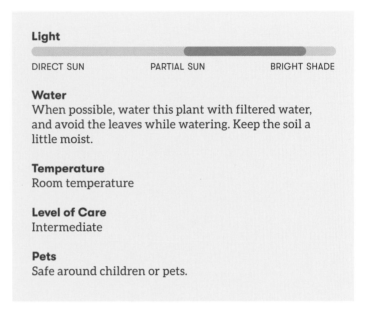

Light

DIRECT SUN   PARTIAL SUN   BRIGHT SHADE

Water
When possible, water this plant with filtered water, and avoid the leaves while watering. Keep the soil a little moist.

Temperature
Room temperature

Level of Care
Intermediate

Pets
Safe around children or pets.

These understory plants are native to neotropical regions of the world (Central and South America, including the south of Mexico and the Caribbean) and typically found in hot lowland areas. These plants have a very low tolerance for cold, with even the ambient chill from windows during the winter months cramping their style. Many species of *Calathea*, *Maranta*, and *Stromanthe* are from rainforests and are susceptible to fungal infection that is often mistaken for low humidity or too much salt, so avoid spritzing with these plants with tap water. While this plant requires more maintenance than others, it is not impossible to please. Create an environment with a warm airflow, and water the plant as soon as its soil dries out (if well saturated, these plants can tolerate a less humid environment).

PLANTING JOURNAL

Caring for something that grows so slowly can require
a lot of meditative observation. These journal pages are
designed to help you with that practice. Begin building
a connection with your plant by coming up with its nickname and recording details about its indoor environment
and ideal watering schedule. Over time, write down or
draw any noteworthy observations, such as the number
of leaves or new growth. Keeping a record of your plant's
health and movements will help you recognize its growth
patterns and what it needs to thrive.

PLANT NICKNAME: _____   **COMMON OR LATIN NAME:** _____

DATE BROUGHT HOME: _____   **POT SIZE (DIAMETER):** _____

ENVIRONMENT

Daily level of light:

>5 hours direct sunbeams

2-4 hours sunbeams

<2 hours sunbeams

Ambient light
(no direct sunbeams)

Near a window?

Y / N

Pet friendly?

Y / N

WATERING SCHEDULE

Frequency:

5 days / 7 days

10 days / 15 days

30 days

OR

Keep soil moist

Let soil dry

SOIL & ROOT HEALTH

Repotted on: _____

Repotting frequency:

6 months / 12 months

Fertilizer used:

Y / N

Fertilization frequency:

1 month

3 months

6 months

NOTES & OBSERVATIONS
(track new growth, number of leaves, seasonal changes, leaf loss, and pests)

_____

_____

_____

_____

_____

_____

_____

_____

_____

_____

PLANT NICKNAME: _____ **COMMON OR LATIN NAME:** _____

DATE BROUGHT HOME: _____ **POT SIZE (DIAMETER):** _____

ENVIRONMENT

Daily level of light:

>5 hours direct sunbeams

2-4 hours sunbeams

<2 hours sunbeams

Ambient light
(no direct sunbeams)

Near a window?

Y / N

Pet friendly?

Y / N

WATERING SCHEDULE

Frequency:

5 days / 7 days

10 days / 15 days

30 days

OR

Keep soil moist

Let soil dry

SOIL & ROOT HEALTH

Repotted on: _____

Repotting frequency:

6 months / 12 months

Fertilizer used:

Y / N

Fertilization frequency:

1 month

3 months

6 months

NOTES & OBSERVATIONS

(track new growth, number of leaves, seasonal changes, leaf loss, and pests)

_____

_____

_____

_____

_____

_____

_____

_____

_____

_____

_____

PLANT NICKNAME: _____ **COMMON OR LATIN NAME:** _____

DATE BROUGHT HOME: _____ **POT SIZE (DIAMETER):** _____

ENVIRONMENT

Daily level of light:

>5 hours direct sunbeams

2-4 hours sunbeams

<2 hours sunbeams

Ambient light
(no direct sunbeams)

Near a window?

Y / N

Pet friendly?

Y / N

WATERING SCHEDULE

Frequency:

5 days / 7 days

10 days / 15 days

30 days

OR

Keep soil moist

Let soil dry

SOIL & ROOT HEALTH

Repotted on: _____

Repotting frequency:

6 months / 12 months

Fertilizer used:

Y / N

Fertilization frequency:

1 month

3 months

6 months

NOTES & OBSERVATIONS
(track new growth, number of leaves, seasonal changes, leaf loss, and pests)

_____

_____

_____

_____

_____

_____

_____

_____

_____

_____

PLANT NICKNAME: _____   **COMMON OR LATIN NAME:** _____

DATE BROUGHT HOME: _____   **POT SIZE (DIAMETER):** _____

ENVIRONMENT

Daily level of light:

>5 hours direct sunbeams

2-4 hours sunbeams

<2 hours sunbeams

Ambient light
(no direct sunbeams)

Near a window?

Y / N

Pet friendly?

Y / N

WATERING SCHEDULE

Frequency:

5 days / 7 days

10 days / 15 days

30 days

OR

Keep soil moist

Let soil dry

SOIL & ROOT HEALTH

Repotted on: _____

Repotting frequency:

6 months / 12 months

Fertilizer used:

Y / N

Fertilization frequency:

1 month

3 months

6 months

NOTES & OBSERVATIONS

(track new growth, number of leaves, seasonal changes, leaf loss, and pests)

_____

_____

_____

_____

_____

_____

_____

_____

_____

_____

_____

_____

PLANT NICKNAME: _____   COMMON OR LATIN NAME: _____

DATE BROUGHT HOME: _____   POT SIZE (DIAMETER): _____

ENVIRONMENT

Daily level of light:

>5 hours direct sunbeams

2-4 hours sunbeams

<2 hours sunbeams

Ambient light
(no direct sunbeams)

Near a window?

Y / N

Pet friendly?

Y / N

WATERING SCHEDULE

Frequency:

5 days / 7 days

10 days / 15 days

30 days

OR

Keep soil moist

Let soil dry

SOIL & ROOT HEALTH

Repotted on: _____

Repotting frequency:

6 months / 12 months

Fertilizer used:

Y / N

Fertilization frequency:

1 month

3 months

6 months

NOTES & OBSERVATIONS

(track new growth, number of leaves, seasonal changes, leaf loss, and pests)

_____

_____

_____

_____

_____

_____

_____

_____

_____

_____

PLANT NICKNAME: _____   **COMMON OR LATIN NAME:** _____

DATE BROUGHT HOME: _____   **POT SIZE (DIAMETER):** _____

ENVIRONMENT

Daily level of light:

>5 hours direct sunbeams

2-4 hours sunbeams

<2 hours sunbeams

Ambient light
(no direct sunbeams)

Near a window?

Y / N

Pet friendly?

Y / N

WATERING SCHEDULE

Frequency:

5 days / 7 days

10 days / 15 days

30 days

OR

Keep soil moist

Let soil dry

SOIL & ROOT HEALTH

Repotted on: _____

Repotting frequency:

6 months / 12 months

Fertilizer used:

Y / N

Fertilization frequency:

1 month

3 months

6 months

NOTES & OBSERVATIONS

(track new growth, number of leaves, seasonal changes, leaf loss, and pests)

_____

_____

_____

_____

_____

_____

_____

_____

_____

_____

_____

_____

PLANT NICKNAME: _____   **COMMON OR LATIN NAME:** _____

DATE BROUGHT HOME: _____   **POT SIZE (DIAMETER):** _____

ENVIRONMENT

Daily level of light:

>5 hours direct sunbeams

2-4 hours sunbeams

<2 hours sunbeams

Ambient light
(no direct sunbeams)

Near a window?

Y / N

Pet friendly?

Y / N

WATERING SCHEDULE

Frequency:

5 days / 7 days

10 days / 15 days

30 days

OR

Keep soil moist

Let soil dry

SOIL & ROOT HEALTH

Repotted on: _____

Repotting frequency:

6 months / 12 months

Fertilizer used:

Y / N

Fertilization frequency:

1 month

3 months

6 months

NOTES & OBSERVATIONS
(track new growth, number of leaves, seasonal changes, leaf loss, and pests)

_____

_____

_____

_____

_____

_____

_____

_____

_____

_____

_____

_____

PLANT NICKNAME: _____   **COMMON OR LATIN NAME:** _____

DATE BROUGHT HOME: _____   **POT SIZE (DIAMETER):** _____

ENVIRONMENT

Daily level of light:

>5 hours direct sunbeams

2-4 hours sunbeams

<2 hours sunbeams

Ambient light
(no direct sunbeams)

Near a window?

Y / N

Pet friendly?

Y / N

WATERING SCHEDULE

Frequency:

5 days / 7 days

10 days / 15 days

30 days

OR

Keep soil moist

Let soil dry

SOIL & ROOT HEALTH

Repotted on: _____

Repotting frequency:

6 months / 12 months

Fertilizer used:

Y / N

Fertilization frequency:

1 month

3 months

6 months

NOTES & OBSERVATIONS
(track new growth, number of leaves, seasonal changes, leaf loss, and pests)

_____

_____

_____

_____

_____

_____

_____

_____

_____

_____

_____

_____

PLANT NICKNAME: _____   **COMMON OR LATIN NAME:** _____

DATE BROUGHT HOME: _____   **POT SIZE (DIAMETER):** _____

ENVIRONMENT

Daily level of light:

>5 hours direct sunbeams

2-4 hours sunbeams

<2 hours sunbeams

Ambient light
(no direct sunbeams)

Near a window?

Y / N

Pet friendly?

Y / N

WATERING SCHEDULE

Frequency:

5 days / 7 days

10 days / 15 days

30 days

OR

Keep soil moist

Let soil dry

SOIL & ROOT HEALTH

Repotted on: _____

Repotting frequency:

6 months / 12 months

Fertilizer used:

Y / N

Fertilization frequency:

1 month

3 months

6 months

NOTES & OBSERVATIONS

(track new growth, number of leaves, seasonal changes, leaf loss, and pests)

_____

_____

_____

_____

_____

_____

_____

_____

_____

_____

_____

PLANT NICKNAME: _____ **COMMON OR LATIN NAME:** _____

DATE BROUGHT HOME: _____ **POT SIZE (DIAMETER):** _____

ENVIRONMENT

Daily level of light:

>5 hours direct sunbeams

2-4 hours sunbeams

<2 hours sunbeams

Ambient light
(no direct sunbeams)

Near a window?

Y / N

Pet friendly?

Y / N

WATERING SCHEDULE

Frequency:

5 days / 7 days

10 days / 15 days

30 days

OR

Keep soil moist

Let soil dry

SOIL & ROOT HEALTH

Repotted on: _____

Repotting frequency:

6 months / 12 months

Fertilizer used:

Y / N

Fertilization frequency:

1 month

3 months

6 months

NOTES & OBSERVATIONS

(track new growth, number of leaves, seasonal changes, leaf loss, and pests)

_____

_____

_____

_____

_____

_____

_____

_____

_____

_____

_____

PLANT NICKNAME: _____   **COMMON OR LATIN NAME:** _____

DATE BROUGHT HOME: _____   **POT SIZE (DIAMETER):** _____

ENVIRONMENT

Daily level of light:

>5 hours direct sunbeams

2-4 hours sunbeams

<2 hours sunbeams

Ambient light
(no direct sunbeams)

Near a window?

Y / N

Pet friendly?

Y / N

WATERING SCHEDULE

Frequency:

5 days / 7 days

10 days / 15 days

30 days

OR

Keep soil moist

Let soil dry

SOIL & ROOT HEALTH

Repotted on: _____

Repotting frequency:

6 months / 12 months

Fertilizer used:

Y / N

Fertilization frequency:

1 month

3 months

6 months

NOTES & OBSERVATIONS
(track new growth, number of leaves, seasonal changes, leaf loss, and pests)

_____

_____

_____

_____

_____

_____

_____

_____

_____

_____

PLANT NICKNAME: _____   **COMMON OR LATIN NAME:** _____

DATE BROUGHT HOME: _____   **POT SIZE (DIAMETER):** _____

ENVIRONMENT

Daily level of light:

>5 hours direct sunbeams

2-4 hours sunbeams

<2 hours sunbeams

Ambient light
(no direct sunbeams)

Near a window?

Y / N

Pet friendly?

Y / N

WATERING SCHEDULE

Frequency:

5 days / 7 days

10 days / 15 days

30 days

OR

Keep soil moist

Let soil dry

SOIL & ROOT HEALTH

Repotted on: _____

Repotting frequency:

6 months / 12 months

Fertilizer used:

Y / N

Fertilization frequency:

1 month

3 months

6 months

NOTES & OBSERVATIONS

(track new growth, number of leaves, seasonal changes, leaf loss, and pests)

_____

_____

_____

_____

_____

_____

_____

_____

_____

_____

_____

PLANT NICKNAME: _____ **COMMON OR LATIN NAME:** _____

DATE BROUGHT HOME: _____ **POT SIZE (DIAMETER):** _____

ENVIRONMENT

Daily level of light:

>5 hours direct sunbeams

2-4 hours sunbeams

<2 hours sunbeams

Ambient light
(no direct sunbeams)

Near a window?

Y / N

Pet friendly?

Y / N

WATERING SCHEDULE

Frequency:

5 days / 7 days

10 days / 15 days

30 days

OR

Keep soil moist

Let soil dry

SOIL & ROOT HEALTH

Repotted on: _____

Repotting frequency:

6 months / 12 months

Fertilizer used:

Y / N

Fertilization frequency:

1 month

3 months

6 months

NOTES & OBSERVATIONS
(track new growth, number of leaves, seasonal changes, leaf loss, and pests)

_____

_____

_____

_____

_____

_____

_____

_____

_____

_____

_____

PLANT NICKNAME: _____

COMMON OR LATIN NAME: _____

DATE BROUGHT HOME: _____

POT SIZE (DIAMETER): _____

ENVIRONMENT

Daily level of light:

>5 hours direct sunbeams

2-4 hours sunbeams

<2 hours sunbeams

Ambient light
(no direct sunbeams)

Near a window?

Y / N

Pet friendly?

Y / N

WATERING SCHEDULE

Frequency:

5 days / 7 days

10 days / 15 days

30 days

OR

Keep soil moist

Let soil dry

SOIL & ROOT HEALTH

Repotted on: _____

Repotting frequency:

6 months / 12 months

Fertilizer used:

Y / N

Fertilization frequency:

1 month

3 months

6 months

NOTES & OBSERVATIONS

(track new growth, number of leaves, seasonal changes, leaf loss, and pests)

_____

_____

_____

_____

_____

_____

_____

_____

_____

_____

PLANT NICKNAME: _____ **COMMON OR LATIN NAME:** _____

DATE BROUGHT HOME: _____ **POT SIZE (DIAMETER):** _____

ENVIRONMENT

Daily level of light:

>5 hours direct sunbeams

2-4 hours sunbeams

<2 hours sunbeams

Ambient light
(no direct sunbeams)

Near a window?

Y / N

Pet friendly?

Y / N

WATERING SCHEDULE

Frequency:

5 days / 7 days

10 days / 15 days

30 days

OR

Keep soil moist

Let soil dry

SOIL & ROOT HEALTH

Repotted on: _____

Repotting frequency:

6 months / 12 months

Fertilizer used:

Y / N

Fertilization frequency:

1 month

3 months

6 months

NOTES & OBSERVATIONS

(track new growth, number of leaves, seasonal changes, leaf loss, and pests)

_____

_____

_____

_____

_____

_____

_____

_____

_____

_____

_____

PLANT NICKNAME: _____ **COMMON OR LATIN NAME:** _____

DATE BROUGHT HOME: _____ **POT SIZE (DIAMETER):** _____

ENVIRONMENT

Daily level of light:

>5 hours direct sunbeams

2-4 hours sunbeams

<2 hours sunbeams

Ambient light
(no direct sunbeams)

Near a window?

Y / N

Pet friendly?

Y / N

WATERING SCHEDULE

Frequency:

5 days / 7 days

10 days / 15 days

30 days

OR

Keep soil moist

Let soil dry

SOIL & ROOT HEALTH

Repotted on: _____

Repotting frequency:

6 months / 12 months

Fertilizer used:

Y / N

Fertilization frequency:

1 month

3 months

6 months

NOTES & OBSERVATIONS
(track new growth, number of leaves, seasonal changes, leaf loss, and pests)

_____

_____

_____

_____

_____

_____

_____

_____

_____

_____

PLANT NICKNAME: _____   **COMMON OR LATIN NAME:** _____

DATE BROUGHT HOME: _____   **POT SIZE (DIAMETER):** _____

ENVIRONMENT

Daily level of light:

>5 hours direct sunbeams

2-4 hours sunbeams

<2 hours sunbeams

Ambient light
(no direct sunbeams)

Near a window?

Y / N

Pet friendly?

Y / N

WATERING SCHEDULE

Frequency:

5 days / 7 days

10 days / 15 days

30 days

OR

Keep soil moist

Let soil dry

SOIL & ROOT HEALTH

Repotted on: _____

Repotting frequency:

6 months / 12 months

Fertilizer used:

Y / N

Fertilization frequency:

1 month

3 months

6 months

NOTES & OBSERVATIONS

(track new growth, number of leaves, seasonal changes, leaf loss, and pests)

_____

_____

_____

_____

_____

_____

_____

_____

_____

_____

_____

PLANT NICKNAME: _____   **COMMON OR LATIN NAME:** _____

DATE BROUGHT HOME: _____   **POT SIZE (DIAMETER):** _____

ENVIRONMENT

Daily level of light:

>5 hours direct sunbeams

2-4 hours sunbeams

<2 hours sunbeams

Ambient light
(no direct sunbeams)

Near a window?

Y / N

Pet friendly?

Y / N

WATERING SCHEDULE

Frequency:

5 days / 7 days

10 days / 15 days

30 days

OR

Keep soil moist

Let soil dry

SOIL & ROOT HEALTH

Repotted on: _____

Repotting frequency:

6 months / 12 months

Fertilizer used:

Y / N

Fertilization frequency:

1 month

3 months

6 months

NOTES & OBSERVATIONS

(track new growth, number of leaves, seasonal changes, leaf loss, and pests)

_____

_____

_____

_____

_____

_____

_____

_____

_____

_____

_____

GLOSSARY

ABIOTIC – Characterized by the absence of life or living organisms.

APICAL DOMINANCE – The phenomenon in which the top bud suppresses the lateral buds of a growing plant shoot.

AROIDS – The collective name for all plants in the Araceae family, including pothos, Monstera plants, and ZZ plants. All aroids are unique, with peace-lily-like flowers covering a spadix, or flower spike, which contains the reproductive parts.

BIOCHEMISTRY – The scientific study of living matter via chemistry.

BIOFILM – Colonies of bacteria and other microorganisms, such as yeast and fungi, that adhere to surfaces that are regularly in contact with water. Algae commonly forms biofilm on rocks at the water's edge of lakes, streams, and such.

BIOPHILIA – The love of the living world, or human's affinity for nature (as opposed to biophobia).

BOTANY – The scientific study of all plant organisms.

COLORATION – Nuclear-derived genetic patterns on leaves that are identical (or near identical) from leaf to leaf.

COLOR RENDERING INDEX (CRI) – The measurement of how colors appear under a light source, as compared to the light of the sun. A CRI of 100 is equivalent to sunlight, and lesser measurements are further from the quality of sunlight.

ETIOLATION – Plants etiolate, meaning their cells and stems elongate and stretch, when there is not enough light. Etiolation is characterized by long, weak stems, small leaves, and pale yellow discoloration of the plant.

GERMINATION – The process of seeds sprouting. Certain plants have germination requirements, such as specific temperatures or light levels.

GYMNOSPERMS – Seed-bearing plants, such as pines, conifers, and ginkgos, do not produce flowers but rather cones.

LICHENS – One of the most ancient forms of quasi-plant-lifes and an alga. Interestingly, the alga can live without the fungus, but the fungus is dependent on the alga.

LUMENS – The measurement of the quantity of light. Also known as light intensity.

MYCELIAL NETWORKS – Fungal bodies in filamentous form in the soil that connect different plants underground and allow them to communicate with each another.

NODES – The part of a stem that normally bears a leaf and aids in propagating. Leaves, flowers, branching stems, and even roots can form from the buds in a node, depending on the plant and its hormones.

ORGANELLES – Cellular organs that store food, produce energy, and discharge waste.

ORGANIC COMPOUNDS – The basis of soils and plant life. Includes decaying plant matter, soil organisms, and carbon compounds (such as sugar, starches, and proteins).

PHLOEM – The living tissue in plants that transports sugars from the leaves to feed the roots.

PHOTOSYNTHESIS – The process in which plants consume light energy to catalyze metabolic reactions to create sugars.

PHYTOCHROME – Photo-receptors that help plants detect light. They are function-ally akin to the cones and rods in the human retina.

PLANT SPECIES – A naturally occurring genetically distinct lineage of distinct plants that share common characteristics and are capa-ble of reproducing.

PLANTCESTORS – Ancient lineages of plants that are the ancestors of modern-day plants. Most of these plants became extinct dduring the Permian–Triassic extinction event, which occurred more than 250 million years ago, but a few live on as unchanged living fossils.

PLANT KINGDOM – The formal taxonomic unit within which all plants are classified.

PLASTIDS – An organelle of plant cells that is responsi-ble for manufacturing and storing food, as well as for coloration, variegation, and photosynthesis.

PROPAGATION – The process by which plants are repro-duced by humans, typically created via asexual methods, such as cuttings, tissue cultures, and divisions.

RULE OF THIRDS – A horticul-tural rule that states that the plant should be two-thirds aboveground and one-third belowground in height. Deviations from the ratio could cause leaf drop or stunted growth.

RHIZOMES – Continuously growing underground plant stem.

SPORTS – Random genetically distinct mutant branches from the parent plant (e.g., a var-iegated vine coming off of a nonvariegated parent vine).

TRANSPIRATION – The natural process of water movement through a plant, with water evaporating from the leaves, stems, and flowers to create water vapor.

TURGOR PRESSURE – The hydraulic pressure that keeps a plant upright! Each plant cell manages its own turgor pressure, all contributing to the stiffness, or droopiness, of your plant.

VARIEGATION – Plastid-derived mutations that create unique patterns on the plant (e.g., the marbling on a pothos or *Monstera*). No two patterns are alike.

VASCULAR SYSTEM – Akin to the blood and lymph systems of humans, a vascular system in plants transports metabolites via the xylem and phloem.

XYLEM – The living tissue in plants that pushes water and nutrients from the roots to the shoots of plants.

NOTES

INTRODUCTION

1. "The guide nods and replies with downcast eyes. 'Yes, I have learned the names of all the bushes, but I have yet to learn their songs. I was teaching the names and ignoring the songs.'" Robin Wall Kimmerer, *Braiding Sweetgrass: Indigenous Wisdom, Scientific Knowledge and the Teachings of Plants* (Minneapolis: Milkweed Editions, 2013), 43.

CHAPTER ONE

1. Alessandra Viola and Stefano Mancuso, *Brilliant Green* (Washington D.C.: Island Press, 2015).
2. Klaus Seeland and Mihir K. Jena, "Knowledge Systems: Indigenous Knowledge of Trees and Forests," in *Encyclopedia of the History of Science, Technology, and Medicine in Non-Western Cultures* (Dordrecht: Springer Netherlands, 2008), 1195.
3. Rene J. Herrera and Ralph Garcia-Bertrand, "The Agricultural Revolutions," in *Ancestral DNA, Human Origins, and Migrations*, ed. Samuel Bowles and Jung-Kyoo Choi (Cambridge: Academic Press, 2018), 475–509.
4. Samuel Bowles and Jung-Kyoo Choi, "The Neolithic Agricultural Revolution and the Origins of Private Property," *Journal of Political Economy* 127, no. 5, 2186–228, accessed December 12, 2020, https://doi.org/10.1086/701789.
5. Jules Janick, *Horticultural Reviews* (Hoboken, NJ: John Wiley & Sons, 2010), doi:10.1002/9780470593776.
6. Alfred Byrd Graf and Roy Perrott, "Houseplant," *Britannica*, accessed December 12, 2020.
7. Sir Hugh Platt, *The Garden of Eden Or, An Accurate Description of All Flowers and Fruits Now Growing in England, with Particular Rules how to Advance Their Nature and Growth* (London: William Leake, 1659).
8. "Potted history of houseplants in our houses and collections," ed Jacq Barber, National Trust, accessed December 12, 2020, https://www.nationaltrust.org.uk/features/potted-history-of-houseplants-in-our-houses-and-collections.
9. Amy Abate, "History of Environmental (Ornamental) Horticulture in the United States," *IR-4 Project*, accessed March 24, 2021, https://www.ir4project.org/ehc/history-of-environmental-ornamental-horticulture-in-the-united-states/#:~:text=1737%20%E2%80%93%20The%20Nursery%20Industry%20is,Linnaean%20Botanic%20Garden%20and%20Nurseries.&text=During%20this%20period%20of%20time,the%20U.S.%20were%20native%20plants.
10. Tom Vilsack, "Secretary's Column: 'The Peoples' Department: 150 Years of USDA,'" *US Department of Agriculture*, accessed March 24, 2021, https://www.usda.gov/media/blog/2012/05/11/secretarys-column-peoples-department-150-years-usda.
11. Tom Knight, "Understanding Plant Names," *Our Houseplants*, accessed March 24, 2021, https://www.ourhouseplants.com/guides/understanding-latin-plant-names.

CHAPTER TWO

1. Edward O. Wilson, *The Biophilia Hypothesis*, ed. Edward O. Wilson and Stephen R. Kellert (Washington, D.C.: Island Press, 1993).
2. Robin Wall Kimmerer, *Braiding Sweetgrass: Indigenous Wisdom, Scientific Knowledge and the Teachings of Plants* (Minneapolis: Milkweed Editions, 2013).

3. Carol J. Adams and Lori Gruen, eds., *Ecofeminism: Feminist Intersections with Other Animals and the Earth* (New York: Bloomsbury, 2014).
4. Claire Jean Kim, *Dangerous Crossings: Race, Species, and Nature in a Multicultural Age* (Cambridge: Cambridge University Press, 2015).
5. Ariel Salleh, *Ecofeminism As Politics: Nature, Marx and the Postmodern* (London: Zed Books, 1997).

CHAPTER FOUR

1. Margaret Bezrutczyk et al., "Sugar Flux and Signaling in Plant-microbe Interactions," *Plant Journal* 93, no. 4 (2018), https://doi.org/10.1111/tpj.13775.
2. James D. Mauseth, *Botany: An Introduction to Plant Biology* (Sudbury, MA: Jones and Bartlett Learning, 2003), 422–27.
3. Christian Fankhauser, "The Phytochromes, a Family of Red/Far-red Absorbing Photoreceptors," *Journal of Biological Chemistry* 276, no. 15 (2001): 11453–56, https://doi.org/10.1074/jbc.R100006200.
4. Esther Hogeveen van Echtelt, "LED: the New Fast-track to Growth: Recipe Development and Practical Applications in Horticulture—Part 1: Global Examples LED Recipes and Development," American Society for Horticultural Science, accessed November 29, 2020, https://ashs.confex.com/ashs/2014/webprogram/Paper20447.html.
5. "What is Color Rendering Index?" *Rensselaer Polytechnic Institute* 8, no. 1 (2004), https://www.lrc.rpi.edu/programs/nlpip/lightinganswers/lightsources/whatisColorRenderingIndex.asp#.
6. "North Carolina Extension Gardener Handbook: Diseases and Disorders," North Carolina Cooperative Extension,

accessed December 12, 2020, https://content.ces.ncsu.edu/extension-gardener-handbook/5-diseases-and-disorders#section_heading_7612.

7. Catriona Mortimer-Sandilands and Bruce Erickson, eds, *Queer Ecologies: Sex, Nature, Politics, Desire* (Bloomington: Indiana University Press, 2010).

8. Jeff Hahn and Julie Weisenhorn, "Managing Insects on Indoor Plants," University of Minnesota, accessed December 12, 2020, https://extension.umn.edu/product-and-houseplant-pests/insects-indoor-plants.

9. Janet McLeod Scott and Joey Williamson, "Common Houseplant Insects and Related Pests," Clemson University Cooperative Extension Home and Garden Information Center, accessed December 12, 2020, https://hgic.clemson.edu/factsheet/common-houseplant-insects-related-pests/.

10. Janet McLeod Scott, "Common Houseplant Insects and Related Pests," Clemson University Cooperative Extension, accessed March 24, 2021, https://hgic.clemson.edu/factsheet/common-houseplant-insects-related-pests

11. W. S. Cransaw and R. A. Cloyd, "Fungus Gnats as Houseplant and Indoor Pests," Colorado State University Extension, accessed December 12, 2020, https://extension.colostate.edu/topic-areas/insects/fungus-gnats-as-houseplant-and-indoor-pests-5-584/.

12. J. A. Bethke and S. H. Dreistadt, "Pest Notes: Fungus Gnats," University of California Agriculture & Natural Resources, accessed December 12, 2020, http://ipm.ucanr.edu/PMG/PESTNOTES/pn7448.html.

13. Lance S. Osborne and W. Chris Fooshee, "Fungus Gnats," Mid-Florida Research and Education Center, accessed December 12, 2020, https://mrec.ifas.ufl.edu/Foliage/entomol/FUNGNAT/fungnat.htm.

14. Hahn and Weisenhorn, "Managing Insects on Indoor Plants."

15. M. L. Flint, "Pests in Gardens and Landscapes: Aphids," University of California Agriculture & Natural Resources, accessed December 12, 2020, http://ipm.ucanr.edu/PMG/PESTNOTES/pn7404.html.

16. "Pests in Gardens and Landscapes: Mealybugs," University of California Agriculture & Natural Resources, accessed December 12, 2020, http://ipm.ucanr.edu/QT/mealybugscard.html.

17. Flint, "Pests in Gardens and Landscapes: Aphids," University of California Agriculture & Natural Resources, accessed December 12, 2020, http://ipm.ucanr.edu/PMG/PESTNOTES/pn 7404.html.

18. "Pests in Gardens and Landscapes: Scales," University of California Agriculture & Natural Resources, accessed December 12, 2020, http://ipm.ucanr.edu/QT/scalescard.html.

19. McLeod Scott, "Common Houseplant Insects and Related Pests."

20. J. A. Bethke, S. H. Dreistadt, and L. G. Varela, "Pests Notes: Thrips," University of California Agriculture & Natural Resources, accessed December 12, 2020, http://ipm.ucanr.edu/PMG/PESTNOTES/pn7429.html.

21. "Pests in Gardens and Landscapes: Thrips," University of California Agriculture & Natural Resources, accessed December 12, 2020, http://ipm.ucanr.edu/QT/thripscard.html.

22. W. S. Cranshaw, "Insect Control: Soaps and Detergents: Fact Sheet," Colorado State University Extension, accessed December 12, 2020. https://extension.colostate.edu/topic-areas/insects/insect-control-soaps-and-detergents-5-547/.

ACKNOWLEDGMENTS

In May of 2020, while I was navigating the fresh uncertainties of the world and my small business, I was approached by Stephanie Holstein and Abby Bussel from Princeton Architectural Press who asked me if I had ever thought about writing a book. They'd read every blog post on our website and thought that our brand had a soul. I can now attest that there couldn't have been a better partner for me—and for Horti—to realize this vision with. I am so thankful to every team member I have interacted with at Princeton Architectural Press for their support, encouragement, and trust, and I can confidently say that they are a publishing house with a soul.

There are two individuals who deserve a special mention for helping to bring this book to life. I'm eternally grateful for my business partner, Bryana Sortino, who took a lot of work off my plate so that I could focus on writing. She has been a powerhouse, managing our team at Horti over the last year. I also want to thank my dearest friend, Leah Kirts, a superb ecofeminist writer who spent countless hours reading my manuscripts and suggesting meaningful edits.

Big high fives to my creative counterparts, Cayla Zahoran and Travis DeMello. We did it! I credit this book's vivid colors and clean lines to the beautiful aesthetic choices they made.

My resident plant doctor, Chris Satch, worked tirelessly on this project. I am in awe of his research capabilities and thankful for his patience in teaching me about all the aspects of plant history and biology that I was unfamiliar with.

The book's foreword was penned by fellow plant lovers Erin Harding and Morgan Doane, who have both supported Horti individually and through their @houseplantclub since its conception.

My heartfelt thanks to Phoebe Cheong, Christopher Griffin, Bryana Sortino, and Amelia Fieldhouse for opening up their beautiful plant-filled homes, which grace many of the pages of this book.

To Sarah Zabrodski, my partner, thank you for being my rock and my compass, helping me to reorient in moments of self-doubt, about this project and life in general.

Thank you to my family, especially my niece and nephew, Ayaan and Ariana, who are taking so much interest in plants. I'm also grateful for my adopted Brooklyn family and friends for trusting my vision with Horti, and for supporting this experiment emotionally and with real dollars: Alexandra Breines, Amar Singh, Blake Johnson, Brett Spiegel, Jeff Greenberg, Jon Wu, Joyce Hanlon, Justin Rancourt, Kareem Amin, Katie Gemmill, Lauren Bugeja, Nicholas Chirls, Nicola Korzenko, Nicolae Rusan, Russell Daiber, Staton Piercey, Scott Miller, Tony Pham, and many more names that I don't have room to list here.

I'd also like to mention these special people who helped shape my life and without whom I wouldn't be where I am today: Anirban Sen, Deepak Bagga, Man Singh Maan, Noah Glass, Theresa J. VanderMeer, and Viral Pandya.

I'm grateful for Horti's team, and every customer, friend, and stranger who purchased a plant from Horti and gave us a chance to realize this dream. We started Horti for our friends and are thrilled to watch our circle keep growing. I have deep gratitude to the early adopters and our subscribers who trust us to curate a plant-filled journey for their homes each month.

And finally, to the readers who have supported this project, thank you for trusting me. It was important for me not to add noise to an increasingly cluttered world of plant care, but to create something I can personally be proud of and that helps you cultivate joy.

Published by
Princeton Architectural Press
202 Warren Street
Hudson, New York 12534
www.papress.com

ISBN 978-1-64896-061-1

Production Editor: Kristen Hewitt
Designer: Natalie Snodgrass

Library of Congress Control Number: 2021947093

IMAGE CREDITS
Page 21: Eldhraun Lava Field, Iceland, 2017.
Photo Jimbo Chan
Page 22: Trumpet lichen, Switzerland, 2020.
Photo Andreas
Page 24: Moss growing on a tree branch,
United Kingdom, 2009.
Photo Graham Wallis
Page 25: *Lepidodendron* and *Sigillaria*, ancient
relatives of clubmosses. Gilbert M. Smith,
Cryptogamic Botany, 1955. Stanford, CA: Stanford
University Press.